SpringerBriefs in Agriculture

T0202866

More information about this series at http://www.springer.com/series/10183

Margaret A. Oliver · Richard Webster

Basic Steps in Geostatistics: The Variogram and Kriging

 Springer

Margaret A. Oliver
Soil Research Centre
  University of Reading
Reading
UK

Richard Webster
Rothamsted Research
Harpenden
UK

ISSN 2211-808X   ISSN 2211-8098   (electronic)
SpringerBriefs in Agriculture
ISBN 978-3-319-15864-8   ISBN 978-3-319-15865-5   (eBook)
DOI 10.1007/978-3-319-15865-5

Library of Congress Control Number: 2015932443

Springer Cham Heidelberg New York Dordrecht London

Printed on acid-free paper

Springer International Publishing AG Switzerland is part of Springer Science+Business Media
(www.springer.com)

Preface

This Brief takes readers, in particular environmental scientists, through the important steps of a geostatistical analysis. Most properties of the environment, such as rainfall, plant nutrients in the soil and pollutants in the air, are measured effectively at points between which there are large gaps. The environment is continuous, however, and environmental scientists and their clients typically want to know the values of those properties between the points, in the gaps; they want to predict in a spatial sense from their data, taking into account the locations of their observations. Geostatistics comprises a set of tools that enable them to do that optimally by methods established for properties that appear to vary randomly in one, two or three dimensions. The variogram is the central tool of geostatistics. It enables scientists to assess whether their data are spatially correlated and to what extent. With a suitable model for it they can combine it with their data to predict by kriging, which in its simpler forms is one of weighted averaging. Kriging is an optimal method of prediction in that it provides unbiased estimates with minimum variance. The technique is now available in many statistical packages so that users can apply it at the press of a few buttons without any idea of what their experimental variogram is like or whether an appropriate model has been fitted to it. We warn against the practice.

We take readers through the stages of computing a reliable experimental variogram from sufficient data and the fitting of suitable mathematical models. These are the most important stages of a geostatistical analysis. If they are done with proper care kriging provides the best possible predictions from data.

Chapter 1 introduces the background to geostatistics with examples of the breadth of current applications. Almost all statistical analyses of environmental data, including geostatistics, depend on sample data, and in this chapter we introduce the basic concepts of sampling: the specification of variables, the support and suitable sampling designs. Chapter 2 describes random variables and regionalized variable theory briefly; this is the theory that underpins geostatistics. For readers who wish to know more of the theory we recommend further reading. Chapter 3 explains how to compute the experimental variogram from regular and irregular sampling designs, the factors that affect the reliability of the variogram and how to model the experimental values reliably. Most users of geostatistics are eager to obtain maps of the variables that interest them, and Chap. 4 illustrates how to do this optimally with kriging. The theory of ordinary kriging is described, and we show how the choice of model for the variogram affects the kriging weights. The way the weights are obtained in kriging makes the method different from other interpolators. In Chap. 5 we return to sampling to meet the needs of spatial analysis. We consider the situation where nothing is known about the scale or pattern of spatial variation and for which a nested survey and analysis provide a solution. We suggest other ways of determining the spatial scale to sample for mapping that use variograms from existing data, either of the variables of interest or of intensive ancillary data, such as those from satellites or aerial photographs. Finally, in Chap. 6 we explore the difficulties that spatial trend poses for geostatistical analysis and how they can be overcome by residual maximum likelihood estimation of the variogram and universal kriging.

Acknowledgments

We thank Dr. A. Vásquez for the data from his survey of trace metals in soil near Madrid for the data on cadmium used for Fig. 3.7 and Dr. B.P. Marchant for fitting the REML variogram (Fig. 6.3) with his own Matlab program.

Contents

Chapter 1
Introduction

Abstract Geostatistics, developed originally in the mining industry from the 1950s onwards, is now being applied widely in environmental science for mapping, monitoring and management. It is based on the theory of random spatial processes. There are numerous examples in soil science, meteorology, agronomy, hydrology, ecology and some aspects of marine science. By taking into account and modelling spatial correlation, geostatistics provides unbiased predictions of environmental variables with minimum and known variance in ways that no other method does. The general technique of prediction is known as kriging. It requires a mathematical model to describe the spatial covariance, usually expressed as a variogram, which in its parameterized form has become the central tool of geostatistics. Successful kriging and estimation of the variogram depend on sampling adequately without bias and with suitable spatial configurations and supports. These differ somewhat from design-based estimation with its emphasis on random sampling.

Keyword Geostatistics · Environmental sciences · Mapping · Random processes · Autocorrelation · Variogram · Kriging · Sampling

1.1 Background to Geostatistics

Geostatistics, developed originally in the mining industry, is now applied widely in the environmental sciences—on the land, in the atmosphere and at sea. Environmental scientists want to make maps of the properties that interest them for more or less large continuous areas over which they often have only sparse sample data at finite numbers of places on small supports. In these circumstances the best that they can do is to estimate or predict in a spatial sense values between sample points. They can use geostatistics to predict at points and over larger blocks in one, two and three dimensions. The techniques have proved valuable, particularly in our own field of soil science for mapping properties such as the concentrations of plant nutrients and potentially toxic chemicals in the soil with known confidence. Their value is general, however, and they are applicable almost whenever there is a desire

© The Author(s) 2015
M.A. Oliver and R. Webster, *Basic Steps in Geostatistics: The Variogram and Kriging*, SpringerBriefs in Agriculture, DOI 10.1007/978-3-319-15865-5_1

to predict and map an environmental variable from sparse data such as in hydrology, geology, petroleum engineering, agriculture, fisheries, meteorology, remote sensing and public health.

Numerous techniques have been proposed over the last 100 years for interpolating from sample information. Some depend on the stratification of regions into discrete classes and prediction from data within those classes. The technique was standard practice in geology and soil survey. Thiessen polygons (Voronoi polygons, Dirichlet tiles), originally proposed for interpolating weather from isolated stations, also depend on a discretization of space and prediction from the stations within the polygons Thiessen (1911). At the other extreme are models of smooth variation in which data at sample points are assumed to lie on or close to mathematically defined surfaces. These include splines, and local and global forms of trend surface analysis. With the availability of such tools, why, you might ask, do we need geostatistics?

The mathematical models are based on assumptions of deterministic relations between the sample locations and the variables of interest. No model can describe in full the variation in the natural world, and any technique for interpolation or spatial prediction will produce results that are more or less in error. It is here that geostatistics has its role because it can provide estimates of those errors. It too is based on an underlying model, but one that incorporates uncertainty. What has made geostatistics especially attractive is that its predictions are in principle unbiased and with minimum and known variance or error. No other technique can match it in general.

Geostatistics comprises a body of statistical technique based on the theory of spatial random processes (see Chap. 2). This is the principal reason why geostatistics has found application in so many fields. It allows us to deal with properties that vary in ways that are far from systematic and at all spatial scales. Geostatistics is now established in the earth sciences as the 'Theory of Regionalized Variables' and due largely to Matheron (1963, 1965). Matheron and co-workers developed it originally for mining, following the empirical work by Krige (1951) for estimating the grades of ore from drill cores in the gold mines of South Africa.

Geostatistics is based on a model of spatially correlated random variation, and estimating the spatial autocorrelation is the first step in a geostatistical analysis. The autocorrelation or variogram functions (see Chap. 2), which can be approximated from sample data, can be described by fairly simple mathematical functions. The parameters of these functions lead to the second important step of geostatistical prediction known as kriging.

1.2 Applications of Geostatistics

1.2.1 Mining and Engineering

As mentioned above, the first applications of geostatistics were in gold mining by Daniel Krige, a mining engineer in the gold fields of the Witwatersrand in South Africa. The motivation was largely profit: are the local concentrations of metals in ores sufficient to make extraction worthwhile? Further, managers and planners, knowing the errors, can balance the financial value of the information against the costs of obtaining it. Geostatistics is also applied to predict diamonds in kimberlites. Bush (2010) describes the use of mixed support kriging to optimize the prediction of diamond concentrations from all available data on different sizes of support. Geostatistics now plays a wider role in mining with matters such as the flow of methane in coal mines and the disposal of waste and its subsequent fate.

The organization, Application of Computers and Operational Research in the Mineral Industry (APCOM), holds frequent conferences that include accounts of geostatistical applications in mining. These are reported in the published proceedings of the organization.

Perhaps the most telling application of geostatistics in civil engineering is that in the design of the Channel Tunnel, the railway tunnel between England and France (Blanchin and Chilès 1993; Chilès and Delfiner 2012). The engineers wanted to bore three tunnels in parallel through the Cretaceous Chalk Marl, a stratum of soft chalky clay. Chilès and colleagues at the Bureau de Recherches Géologiques et Minières (BRGM) analysed the depths of the top and bottom of the stratum recorded from bore holes and mapped the configuration of the Marl and its upper and lower bounds. As the tunnels were bored, the engineers reported the discrepancies between the kriged predictions of the upper and lower limits of the stratum and what they encountered refined their predictions with the new information. Chilès and Delfiner conclude: '… the observations were generally in good agreement with the geostatistical model and its predicted accuracy.' The whole operation was a huge success.

1.2.2 Environmental Pollution

In the last couple of decades scientists in research institutes and universities have harnessed geostatistics to estimate and map potentially toxic substances in the environment and to identify sources of pollution. In 2008 Soares (2010), in reviewing that progress, was concerned that governments and industry had been slow to recognize it and to implement the technology, but the situation is changing. Governments world-wide are recognizing the damage caused by the pollution of water, soil and air, and the risks to public health. Geostatistics is playing an ever-increasing role in delineating zones affected and the degree to which they are affected. The literature of the last 3 years is rich in reports of case studies.

A search of the SCI data base with key words 'pollution' and 'geostatistics' revealed more than 600 papers, of which at least 100 have been published since the beginning of 2012. Refining the search by the addition of 'environment' still listed more than 340 papers describing the outcomes of geostatistical analyses of survey data. We can mention but a few of them.

A paper by Komnitsas and Modis (2009) exemplifies the effects of mining. Several million cubic metres of waste from coal mines have been spread in an area of 39 km^2 some 200 km south of Moscow, Russia. As a result concentrations of arsenic (As), zinc (Zn), nickel (Ni) and chromium (Cr) in the soil widely exceed local tolerable thresholds. The authors mapped the risks they pose. Similarly, Zhong et al. (2014) mapped the distributions of As, Cr, cadmium (Cd), lead (Pb) and mercury (Hg) in the soil of an area of 33 500 km^2 of the Yunnan–Guizhou plateau in southern China where non-ferrous metals have been mined for hundreds of years. Their maps show the close association between large and potentially toxic concentrations of As, Cd, Hg and Pb and metallurgical plants from which the elements have been distributed. Chromium, in contrast, seems to derive locally from the rocks. An unusual application of geostatistics to pollution is that of Meerschman et al. (2011). They sampled the soil around Ypres, Belgium, where a battle was fought during the First World War. They identified 'hot spots' of copper (Cu), Pb and Zn where concentrations of the metals, the residues from ordnance, exceeded the national sanitation thresholds for soil.

Toxic concentrations of elements in some regions are entirely natural. For example, arsenic (As) concentrations in the ground water in Bangladesh. These have serious consequences for public health because 97 % of the people there draw on the ground water for drinking. Gaus et al. (2003) analysed geostatistically measurements from more than 3 000 boreholes and estimated the probabilities of concentrations in excess of the World Health Organization's guide limit of 10 μg As l^{-1} and the more lenient Bangladesh limit of 50 μg As l^{-1}. The ground water beneath large tracts of the south of the country had concentrations of As that exceeded the above thresholds, and tens of millions of people were at risk of poisoning by drinking the water. In some regions toxic metals in soil derive from both rocks and industry. Sollitto et al. (2009) used factorial kriging to distinguish short-range variation in Zn, Pb, Cd and Ni, which they attributed to local inheritance from the rocks, from long-range variation caused by diffuse pollution and deposition from the air.

In yet other cases investigators have used geostatistics to map distributions of potentially toxic elements in the search for sources of pollution only to discover later that the local concentrations were entirely natural. Atteia et al. (1994), for example, mapped Cd in the topsoil around La Chaux-de-Fonds in the Swiss Jura to search for sources of pollution where previous sampling had revealed extraordinarily large concentrations of the element well in excess of the safety threshold. The result did not match supposed pollution from the town, and only after a series of surveys targeted in the patches of large concentrations did Dubois et al. (2002) find the explanation: the underlying rock in those patches was rich in Cd.

1.2.3 Precision Agriculture (PA)

About 25 years ago agronomists realized that geostatistics could be harnessed for precision agriculture, and they have made substantial progress in the technology since. It is used for the site-specific management of crop nutrients, pH, irrigation, weeds and crop pests (see Oliver 2010, for examples). Infestations of weeds, pests and diseases vary in intensity within arable fields and paddocks. In many instances their distributions are patchy, and farmers want to treat the land where they occur and not where they are absent. The technology to target types of weeds based on kriged maps is developing apace, but there is still some way to go before it becomes standard practice because the costs of the technology exceed the possible savings in herbicides, which are cheap, and gains from increased yields in treated patches. Castrignanò et al. (2012) used cokriging to delineate zones of *Bactrocera olea*, the olive fruit fly, to optimize their monitoring effort. The spatial pattern of the fly changed over the monitoring period, and they were able to recommend different monitoring zones for summer and October because the population density distribution of the fly was spatially structured over large areas and changed over time.

Among the most serious agricultural pests are nematodes; they are thought to cause losses of the order US$100 $\times$ 10^9 per year worldwide. Like weeds, their distributions within fields tend to be patchy, and they have also been analysed geostatistically (see Webster 2010; Hbirkou et al. 2011). Unlike weeds, their distributions cannot be seen or inferred until the pests have done their damage. Any assessment of their presence, their numbers and their spatial distributions must be based on sample counts made before the crops start to grow. Webster and Boag (1992) devised an extreme form of unbalanced nested sampling to identify for modest cost the approximate sizes of patches of the cyst nematodes of the genera *Heterodera* and *Globodera*, and from their results they sampled on a grid with a suitable interval to map the patches by kriging. Evans et al. (2002, 2003) built on that research to estimate, map and control the potato cyst nematodes *Globodera pallida* and *G. rostochiensis* on commercial farms. They calculated the savings that potato growers in Britain could make by fairly dense sampling and not treating with expensive nematicide the uninfested patches.

Ge et al. (2008) used geostatistics to map the spatial variation of several components of the quality of cotton fibre in the USA. The ranges of spatial correlation for the properties were similar as were the patterns of spatial variation, although some properties had inverse relations with one another. The spatial variation of EC_a was also similar to those of the quality of the cotton fibre. Ge et al. (2008) combined the maps of individual fibre properties with the United States Department of Agriculture—Commodity Crop Corporation Loan Schedule for Upland Cotton to create a loan rate map related to fibre quality. There was a loan rate difference of 20 cents kg^{-1} in the field, which can have a large impact on the producer's revenue.

An interesting application of geostatistics to PA was by Araújo e Silva Ferraz et al. (2012) to the variation in force required to detach coffee berries. Coffee growers want to harvest only mature berries from which the best coffee derives. Mature berries detach more readily than green immature ones, and so growers who use machines for harvesting can use kriged maps of detachment force to decide where and when to harvest the mature berries selectively to ensure the quality of their coffee.

1.2.4 Fisheries

Petitgas (1993, 2001) explained to fishery scientists how geostatistical theory and methods might be applied to estimate and map fish stocks in the sea. However, surveys are problematic because the main species of commercial interest are mobile, and so estimates must be up-to-date and based on rapid survey. Abundances of fish can be sensed acoustically, and despite the complex relations between the signals and actual abundance there have been successes. For example, Jardim and Ribeiro (2008) in a study of sampling designs could map the abundance of hake on the continental shelf off the coast of Portugal.

Mapping the abundances of shell fish is much simpler; the individuals do not move far or fast, they are confined to shallow water, and their numbers and sizes can be determined on well-defined supports, typically within quadrats of fixed dimensions. Adams et al. (2010) surveyed two regions of Georges Bank, close to Nantucket Island, USA, annually from 1999 to 2007 inclusive. They used an underwater camera to enable them to count sea scallops, *Placopecten magellanicus*. They analysed their count data geostatistically and mapped their estimates of species density. The authors were able to recommend strategies for sampling Georges Bank, and for managing the fishery in zones identified by their mapping so that the stocks of scallops are never severely depleted anywhere on the Bank.

1.3 Sampling

Above we mention that the environment is continuous and that its properties cannot always be measured everywhere; quantitative information about such properties can derive only from samples. Much information now derives from either remote or proximal sensors that provide full cover of areas of interest. Such sources may result in too much information that is also often noisy and would benefit from sub-sampling. Sound sampling is crucial therefore. We may require sample information for one of three purposes, or perhaps all:

(a) to estimate average values or total quantities for particular regions,
(b) to predict local values at sites unvisited during survey, or
(c) to sub-sample intensive data from sensors.

We usually want also that the estimates or predictions on average equal the true values, i.e. that they are unbiased. A second requirement is that the estimates or predictions lie within a tolerable error; we want ones in which we can have confidence. For this the sampling must be sound in design and sufficient.

The theory of sampling for estimation was developed during the 1930s and has provided the basis for sampling in many spheres of endeavour. Cochran (1977) draws on that theory to set out the principles of sound design and the statistical analyses that follow from them. We call this the classical approach. Spatial prediction depends on a somewhat different set of principles based on the theory of regionalized variables and leads to different designs and different forms of analysis, and this is what geostatistics is about. De Gruijter et al. (2006) distinguish the two forcefully before going on to describe the methods and the analyses of the data that accrue from sampling. Webster and Lark (2013) develop the ideas further by examining the practical and financial constraints faced by environmental scientists and managers in addition to the technical details.

In the classical approach random selection is essential to guarantee unbiased estimates with known variance. Simple random sampling of the environment is usually inefficient in that large samples are needed to provide adequate confidence. Various kinds of stratification, including classification into types of rock or soil, will usually allow investigators to narrow confidence intervals for a given effort or achieve adequate confidence with smaller samples.

One can also build in prior knowledge to increase the efficiency of designs (see Webster and Lark 2013). De Gruijter et al. (2006) call the whole process 'design-based estimation'.

Geostatistics is also concerned to estimate unknown quantities. These quantities are not the mean values of whole regions; rather they are values at points or in small blocks of land or rock, or bodies of water or air; they are estimates of actual values at unvisited places rather than of unknown parameters of populations. Again, practitioners want their estimates, or predictions as they are usually known to distinguish them from estimates of parameters, to be unbiased on average and to lie within tolerable error. One could use design-based methods, but they would be very inefficient because they would fail to take into account the spatial dependence in data. Instead geostatistics assigns randomness to the environment; it is based on a model in which the actual values of a variable, say z, and which we might want to predict, are the outcomes of random processes. In other words, we have to have in mind a model of the real world that incorporates randomness. What it means in practice is that we need no longer select sites for sampling at random because the randomness is in the model. De Gruijter et al. call the process of prediction 'model-based prediction' for this reason. We shall usually want to avoid bias in the selection, but our principal concern will be to sample enough to enable us to predict accurately throughout the regions of concern, and to make maps, for example. Designing such schemes for sampling requires an understanding of the random spatial model (see Chap. 2) and its practical consequences for prediction.

1.3.1 The Domain

The region to be sampled must be defined. It might be a farm or a field on a farm, a catchment or an administrative district, and it will be bounded geographically. Each of these taken as a whole could constitute the geographic domain.

We have to consider next what part of the material within that region we are to sample. In each case we must define what part of the environment is to be studied. It must lead to a rule or set of rules that can be followed unequivocally and easily and suit the circumstances of the study. This rule or these rules define the total domain, which in classical statistics is known as the 'population' or the 'frame'.

1.3.2 The Variables

The properties of the material in the domain must also be defined. In the farming context, these are likely to be the soil's pH and concentrations of potassium and phosphorus. If we are concerned with air pollution then properties of interest are likely to be the concentration of oxides of nitrogen, NO_x, and particles of diameter smaller than 2.5 μm, the $PM_{2.5}$ of the air.

1.3.3 Units and Support

A sample of a material in the environment comprises a set of *units*, which for statistical purposes we must regard as discrete and on which we make our measurements. To map the distribution of rainfall our units would be rain gauges. In a survey of soil the units might be auger cores, or sets of auger cores from small quadrats and bulked. Agencies monitoring pollution take known volumes of air on which to make their measurements.

The size of these units matters a great deal because the variation that can be observed from measurements on them depends on their size. Some cores of soil taken with an auger of 2 cm diameter are likely to include large root channels and holes made by burrowing animals whilst others do not. Most cores of 80-cm diameter are likely to include such channels, and so measurements of porosity are smoothed. Similarly, bulking soil from small cores from larger quadrats for laboratory analyses leads to a kind of averaging. The support of a rain gauge is strictly the area of its top. In general, the larger are the units the more variation they encompass and the less variation lies between them for us to observe; variation within the units is effectively hidden. We must specify at the outset of a survey the dimensions of the units—not only their size, but also their shape and orientation. This combination is known as the *support* of the sample. It should be held constant throughout a survey, and it should be stated when results are reported.

1.3.4 Practical Matters

We deal in detail with plans for sampling in Chap. 4. There we describe how to choose the positions of the units for mapping and to estimate the crucial intermediary, the variogram or covariance function. Here we mention some of the snags that surveyors almost inevitably encounter.

One of the most common snags is that a point designated by its spatial coordinates for a survey of some attribute of the land falls on a road or in a river. If the aim of the survey is to predict and map some property of the soil then that point must be rejected. Nevertheless, even coverage is desirable for mapping, and one would not want large gaps in it, so one might choose a substitute nearby. Such eventualities should be anticipated, and some rule for substitution should be set beforehand. It might be: choose another point y metres away in direction θ. The distance y should be a small proportion of the average distance between neighbouring points so as to fill the gap.

Less clear is what a surveyor should do if he finds that a predetermined point lies in a gateway or falls in a hedge. In both situations the soil there is likely to be atypical. The points lie within the geographic domain; but are they part of the population to be investigated? Again, the question should have been anticipated and a rule provided.

1.4 The Essence of Geostatistics

In this brief we set out the essentials of geostatistics for environmental applications.

These include a short chapter, Chap. 2, on the theory of spatial random processes and their mathematical representation in terms of covariances and variograms. It is followed by a longer Chap. 3 on the estimation of covariance functions and variograms, and their modelling by permissible functions. We emphasize the need to give attention to the issues raised in this chapter. In Chap. 4 we describe ordinary kriging for predicting values of stationary environmental variables at unobserved places at points or over larger blocks of land without bias and with minimum variance. The parameters of a variogram model are crucial for this, and we show how the models chosen and fitted to sample variograms affect the outcome of kriging. Estimated variograms and kriged predictions depend on the intensity of sample data and their configurations, and sampling for both is the subject of Chap. 5. Finally in Chap. 6 we introduce readers to the problem of trend, i.e. non-stationary processes, and mixed models. The last four chapters are illustrated with data from case studies.

Chapter 2
Regionalized Variable Theory

Abstract Many physical, chemical and biological processes have acted to create the current environment with the result that the variation appears to be random. Practical geostatistics treats the results as if they were the outcomes of correlated random processes and is underpinned by assumptions of stationarity. Variation may be treated as second-order stationary and represented by covariance functions. The somewhat weaker assumption of intrinsic stationarity leads to a more general analysis based on the variogram as a description of the variation. Quasi-stationarity limits stationarity to local areas, and with sufficient data the assumptions can be applied locally. If there is trend then more complex assumptions are needed; these usually comprise a combination of deterministic spatially smooth trend plus random residuals that are spatially correlated and stationary to some degree.

Keywords Random function · Random variable · Regionalized variable · Stationarity · Intrinsic hypothesis · Quasi-stationarity · Trend

2.1 Random Variables and Regionalized Variable Theory

The processes that act in the environment obey the laws of physics, and are in that sense deterministic: the variation we observe has its physical causes. Nevertheless, numerous processes have combined and interacted to produce the current environment, and the results are so complex that the variation appears as though it were generated randomly (Webster 2000). In geostatistical terms, we regard the grades of ores, properties of the soil, or the rainfall of a region, of almost any size, as the realizations of random processes.

Based on this view the value of, for example, a soil property such as its pH, at any place, $\mathbf{x}$ denoting its coordinates in two dimensions, is just one of the infinitely many that are possible there. We associate with each place $\mathbf{x}$ not just one value but a whole suite of values with a mean, a variance and higher-order moments of a distribution. The actual value at $\mathbf{x}$ is regarded as just one value from that distribution, allocated at random. Thus the value of the variable at $\mathbf{x}$ is treated as a random variable, which we denote with the capital Z. The set of random variables for all $\mathbf{x}$ in $\mathfrak{R}$ constitutes a random function, random process or stochastic process. Random variables in the real space, which may be one-, two- or three-dimensional, are also called 'regionalized variables', and hence we have the theory of region- alized variables mentioned above.

A random function has no mathematical description in the way that a deter- ministic one has, i.e. it cannot be written as an equation. Nevertheless, it may have 'structure' in that there is correlation in space, or in time (for signals). This means that values at different places may be related to one another in a statistical sense. Intuitively, we expect the features of the environment at places near to one another to be similar, whereas those at widely separated places are less likely to be. This intuition is formalized in the theory of random functions. We must realise that the randomness is a mental model of the world and not a property of the environment.

2.1.1 Stationarity

Stationarity underpins the practicality of geostatistics; it is an assumption that enables us to treat data as though they have the same degree of variation over a region of interest. We can represent the random process by the model

$$Z(\mathbf{x}) = \mu + \varepsilon(\mathbf{x}), \tag{2.1}$$

where μ is the mean of the process and $\varepsilon(\mathbf{x})$ is a random quantity with a mean of zero and a covariance, $C(\mathbf{h})$, where $\mathbf{h}$ is the separation in space and known as the lag.The covariance is

$$C(\mathbf{h}) = \mathrm{E}[\varepsilon(\mathbf{x})\varepsilon(\mathbf{x}+\mathbf{h})], \tag{2.2}$$

which is equivalent to

$$C(\mathbf{h}) = \mathrm{E}[\{Z(\mathbf{x}) - \mu\}\{Z(\mathbf{x}+\mathbf{h})\} - \mu\}] = \mathrm{E}[\{Z(\mathbf{x})\}\{Z(\mathbf{x}+\mathbf{h})\} - \mu^2]. \tag{2.3}$$

Here $Z(\mathbf{x})$ and $Z(\mathbf{x}+\mathbf{h})$ are the values of the random variable Z at places $\mathbf{x}$ and $\mathbf{x}+\mathbf{h}$ and E denotes the expectation. This covariance depends on $\mathbf{h}$ and only on $\mathbf{h}$, the separation between samples in both distance and direction; it is a function of $\mathbf{h}$. The assumption on which this is based is that of second-order stationarity. In the real world, we often encounter situations in which we cannot assume that the mean is constant, and if so the covariance cannot exist. Such a situation need not be a stumbling block; we can simply weaken the assumption of stationarity to that of what Matheron (1963) called *intrinsic stationarity* in which the expected differences are zero,

$$E[Z(\mathbf{x}) - Z(\mathbf{x} + \mathbf{h})] = 0, \tag{2.4}$$

and the covariance of the residuals is replaced by the variance of the differences to measure the spatial relations:

$$\mathrm{var}[Z(\mathbf{x}) - Z(\mathbf{x} + \mathbf{h})] = E\left[\{Z(\mathbf{x}) - Z(\mathbf{x} + \mathbf{h})\}^2\right] = 2\gamma(\mathbf{h}). \tag{2.5}$$

Here $\gamma(\mathbf{h})$ is the semivariance at lag $\mathbf{h}$, and as a function of $\mathbf{h}$ it is the variogram. The variogram is based on differences, and provided Eq. (2.4) holds locally it is valid. This property makes the variogram more generally useful than the covariance function. In Chap. 3 we describe how to compute the covariance and variogram functions. We focus on the variogram because of its generality and go on in Chap. 3 to describe variogram modelling.

For second-order stationary processes the covariance function and variogram are equivalent:

$$\gamma(\mathbf{h}) = C(0) - C(\mathbf{h}), \tag{2.6}$$

where $C(0) = \sigma^2$ is the variance of the process.

A process that appears stationary at one scale might at another scale appear to embody trend, that is, a systematic component. At this scale we might have to elaborate the simple model represented in Eq. (2.1) by

$$Z(\mathbf{x}) = u(\mathbf{x}) + \varepsilon(\mathbf{x}), \tag{2.7}$$

in which $u(\mathbf{x})$ is a deterministic trend term that replaces the constant mean, μ. Its variogram,

$$\gamma(\mathbf{h}) = \frac{1}{2} E\left[\{\varepsilon(\mathbf{x}) - \varepsilon(\mathbf{x} + \mathbf{h})\}^2\right], \tag{2.8}$$

is no longer the same as

$$\gamma(\mathbf{h}) = \frac{1}{2} E\left[\{Z(\mathbf{x}) - Z(\mathbf{x} + \mathbf{h})\}^2\right], \tag{2.9}$$

of Eq. (1.5). It is the variogram of the residuals from the trend. We explain what to do to estimate the variogram in the presence of trend in Chap. 6.

Chapter 3
The Variogram and Modelling

Abstract Accurate estimates of variograms are needed for reliable prediction by kriging and subsequent mapping and for optimizing sampling schemes. Sample variograms are usually computed by the method of moments at a sequence of lags, and one or more 'authorized' functions are fitted to them. A variogram may be computed along transects or on grids at regular intervals or in bins from irregularly scattered data. Accuracy of the variogram depends on the size of sample, the number of lags at which it is estimated and the lag interval relative to the spatial scale of variation, the marginal distribution of the variable, anisotropy and trend. Robust estimators can deal with extreme values, outliers. Variograms may be bounded (for second-order stationary processes) or unbounded (intrinsically stationary only), and there are few simple authorized functions for modelling them. The parameters of the models summarize the spatial variation and are needed for subsequent kriging. Computing the variogram in at least three directions can identify anisotropy if it is present. Diagnostics including residual mean squares and the Akaike Information Criterion help in the selection of the best fitting model.

Keywords Experimental variogram · Method of moments · Model parameters · Lag interval · Spatial scale · Marginal distribution · Anisotropy · Outliers · Robust estimators · Nugget variance · Sill variance · Model diagnostics

The variogram is the cornerstone of many geostatistical applications. The experimental variogram and any model fitted to it should be accurate. Only then can the model describe the variation reliably. Kriging requires a variogram, and it is by ensuring its accuracy that you will eventually obtain minimum-variance predictions by kriging. If the variogram describes the variation poorly then the kriged predictions are likely to be poor also, and they might have little or no validity no matter how 'pretty' the map. The term 'cartographic pornography' has been used by those who realize that no confidence can be placed in many of the beautiful smooth maps that exist because of sparsity of the data that underlies them (see Chap. 4). Further, the parameters of the variogram model may be used later for sample design and the kriged estimates for decision-making; computing experimental variograms and modelling them should not be treated in a cavalier fashion.

© The Author(s) 2015
M.A. Oliver and R. Webster, *Basic Steps in Geostatistics: The Variogram and Kriging*, SpringerBriefs in Agriculture, DOI 10.1007/978-3-319-15865-5_3

This chapter illustrates the essential steps in obtaining reliable experimental variograms by Matheron's (1965) method-of-moments (MoM) and modelling them. In some geostatistics packages and several GISs, computing the variogram and kriging from the data is automated. As a consequence of such a 'black box' approach, the variogram is computed and modelled, and the parameter values from the model are inserted into the kriging equations without any intervention or assessment by the user. As a result the user has no idea of the variogram's form (it might even be pure nugget) or whether the model is a good fit. There are many other reasons for poor variograms and their models, for example too few data (Webster and Oliver 1992), unsuitable models, poor fitting, faulty processing and misunderstanding. These are matters that form the basis of this chapter. Our aim is to prevent researchers from wasting time on analyses for which their data are unsuitable, and to guide them through the stages that will ensure that their variograms are 'fit-for-purpose'.

3.1 The Experimental Variogram

The first task in turning theory into practice is to estimate the variogram from sample data, say $z(\mathbf{x}_1)$, $z(\mathbf{x}_2)$,..., where $\mathbf{x}_1$, $\mathbf{x}_2$,... denote the positions of the sample in two-dimensional space. We assume that those positions have been selected without bias. They need not be random, as in design-based estimation, because we treat the variables as the outcomes of random processes. Therefore, we can take a relaxed attitude to the sampling design, which may be systematic, random, nested or some combination (see Chap. 5). The usual equation to compute the variogram is Matheron's method of moments (MoM) estimator:

$$\hat{\gamma}(\mathbf{h}) = \frac{1}{2m(\mathbf{h})} \sum_{i=1}^{m(\mathbf{h})} \{z(\mathbf{x}_i) - z(\mathbf{x}_i + \mathbf{h})\}^2, \tag{3.1}$$

where $z(\mathbf{x}_i)$ and $z(\mathbf{x}_i + \mathbf{h})$ are the observed values of z at places $\mathbf{x}_i$ and $\mathbf{x}_i + \mathbf{h}$, and $m(\mathbf{h})$ is the number of paired comparisons at lag $\mathbf{h}$. By changing $\mathbf{h}$ we obtain an ordered set of semivariances; these constitute the experimental or sample variogram. The way that Eq. (3.1) is implemented as an algorithm depends on whether the data are regularly spaced in one dimension, are on a regular grid or are irregularly distributed in two dimensions.

3.1.1 Computing the Variogram from Regular Sampling in One Dimension

Regular sampling in one dimension may be horizontal or vertical (e.g. down boreholes or through the atmosphere) along transects. The lag, $\mathbf{h}$, becomes a scalar

$h = |\mathbf{h}|$ that replaces $\mathbf{h}$ in Eq. (3.1). Semivariances, $\hat{\gamma}(h)$, can be computed only at multiples of the sampling interval. Figure 3.1a shows how the comparisons between pairs of points are made; first for $h = 1$ and then for $h = 2, 3, \ldots$. This results in a set of semivariances, $\hat{\gamma}(1), \hat{\gamma}(2), \hat{\gamma}(3), \ldots$, i.e. a one-dimensional experimental variogram which we can plot as a graph of $\hat{\gamma}(h)$ against h as in Fig. 3.1b. There may be positions along a transect where, for various reasons, there are no observations. These missing data do not present a problem; they simply result in fewer comparisons for Eq. (3.1).

Transects may be aligned in several directions, for example to identify anisotropy when at least three directions should be used (see Sect. 3.2.5). The same procedure may be used to compute these variograms, and Eq. (3.1) will provide a separate set of estimates for each direction.

Fig. 3.1 **a** Comparisons for computing a variogram for three lag intervals from a regular sample every 10 m along a transect and **b** semivariances plotted against the first three lag intervals to form the sample variogram (other possible semivariances shown as *crosses*)

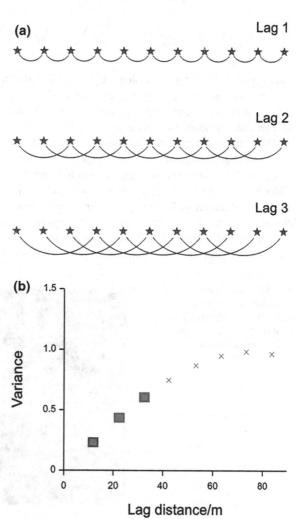

3.1.2 Computing the Variogram from Regular and Irregular Sampling in Two Dimensions

Data from regular grid sampling in two dimensions can be analysed in one of three ways. First, the grid can be treated as a series of transects in two dimensions—it is one way in which you can investigate anisotropy, i.e. directional differences, in the variation. The variogram can be computed as above, but in several directions of the grid separately, for example, along the rows and columns of the grid and on the diagonals. Second, the variogram can be computed in two dimensions as follows:

$$\hat{\gamma}(p,q) = \frac{1}{2(m-p)(n-q)} \sum_{i=1}^{m-p} \sum_{j=1}^{n-q} \{z(i,j) - z(i+p, j+q)\}^2 ,$$

$$\hat{\gamma}(p,-q) = \frac{1}{2(m-p)(n-q)} \sum_{i=1}^{m-p} \sum_{j=q+1}^{n} \{z(i,j) - z(i+p, j-q)\}^2 ,$$

(3.2)

where p and q are the lags along the rows and down the columns of the grid, respectively. In general, the lag interval is that of the grid. The variogram is computed for lags from $-q$ to q and from 0 to p. The output from this is then plotted as a two-dimensional variogram as in Fig. 3.2.

Third, the variogram can be computed over all directions (omnidirectional) for both regular and irregular sampling designs. For a grid the initial nominal lag interval should be that of the grid spacing, whereas for irregularly scattered data the choice is wider because the observations may be separated by potentially unique lags in both distance and direction. Figure 3.3 explains how we can obtain semivariances over all directions in two dimensions by placing the lags into bins. We choose a nominal lag interval in both distance and direction as shown in grey in the figure. The width in distance is designated w, which for irregularly scattered data could be the average

Fig. 3.2 Two-dimensional anisotropic experimental variogram of a simulated field of 100 000 values computed to 11 intervals on the principal axes

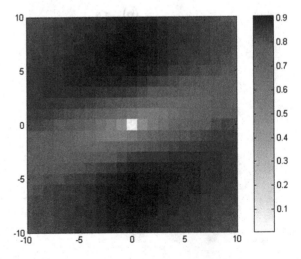

Fig. 3.3 The geometry in
two dimensions for
discretizing the lag into bins
by distance and direction

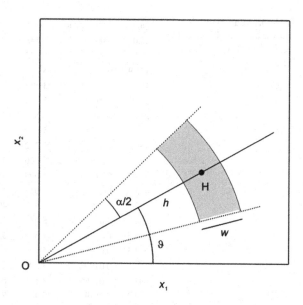

distance between neighbouring sampling points. The angular width is the angle α.
All pairs of comparisons that fall within that bin and contribute to $\hat{\gamma}$ are attributed to
its centroid at H with nominal lag h, ϑ and where ϑ is the lag direction. The lag is
usually incremented in steps of w and ϑ so that each paired comparison falls into one
and only one bin. To compute the omnidirectional variogram, the angular width of
the bins is set to $\alpha = \pi$ (i.e. 180°). We may also compute the variogram in a set of lag
directions, ϑ, (see Sect. 3.3.2).

3.2 Factors Affecting the Reliability of Experimental Variograms

3.2.1 Sample Size

The accuracy of the variogram depends primarily on one's having enough data at a
suitable density or separating interval. It also depends on the design or configuration of the sample because of the way that the variogram is usually computed. The
random function model (see Chap. 2) enables us to have the multiple realizations
required by theory; we treat each comparison between any pair of data as a single
realization of the process. Therefore, for every lag interval we require many
comparisons to ensure reliability of the estimated semivariances. At the shortest
lags or separating distances, we might have rather few paired comparisons for two-
dimensional data. As the lag interval between data increases, however, the number
of comparisons increases (see Table 3.1). At some distance that depends on the
number of data the number of pairs for comparison starts to decrease, although the

Table 3.1 Lag intervals, semivariances and counts for $\log_{10} K^+$ at Broom's Barn Farm

Lag/m	Semivariance	Counts	Lag/m	Semivariance	Counts
48.0	0.00726	1 545	55.3	0.00818	53
92.5	0.00971	2 793	100.1	0.01245	90
131.9	0.01128	3 217	135.7	0.01439	124
169.3	0.01280	3 592	167.7	0.01715	137
212.6	0.01488	5 456	210.3	0.01808	223
253.0	0.01661	4 562	252.9	0.01617	190
293.0	0.01777	5 524	292.2	0.01647	230
334.8	0.01905	5 792	337.2	0.02171	243
374.8	0.01936	5 226	374.6	0.02059	193
414.3	0.01996	5 699	414.6	0.01616	236
453.0	0.01960	4 918	452.0	0.01949	174
492.2	0.02016	5 447	490.8	0.02363	236
534.5	**0.01930**	**5 484**	**534.3**	**0.01786**	**226**
575.6	0.01881	4 513	574.7	0.01706	187
614.5	0.01877	4 189	615.1	0.02181	167
653.5	0.01806	3 524	653.9	0.02363	139
693.2	0.01866	3 572	692.4	0.02228	142
734.9	0.01792	3 136	733.5	0.01774	123
775.1	0.01766	2 689	773.5	0.01832	119

numbers might still be much larger than for the first few lags (Table 3.1). The larger numbers do not imply greater reliability, however, because individual data are used repeatedly, and the estimated semivariances are more or less correlated with one another. As a result, you should not rely on the number of comparisons as a guide to the reliability of your variogram when you have too few data to ensure accuracy.

We illustrate the effect of sample size with data on exchangeable potassium in the topsoil (0–23 cm) from a survey at Broom's Barn Farm (an experimental farm of 80 ha in Suffolk, England), which was first analysed by Webster and McBratney (1987). Table 3.2 summarizes the statistics. There were 434 sampling sites at an interval of 40 m on a square grid. The data were transformed to common logarithms ($\log_{10}$) because the skewness coefficient is 2.04 (see Table 3.2 and an explanation in

Table 3.2 Summary statistics of potassium at Broom's Barn Farm

	$K^+/mg\ l^{-1}$	$\log_{10}(K^+)$	$K^+/mg\ l^{-1}$	$\log_{10}(K^+)$
Number of data	434	434	87	87
Minimum	12.0	1.0792	14.0	1.146
Maximum	96.0	1.9823	70.0	1.845
Mean	26.31	1.3985	26.7	1.404
Median	25.0	1.3979	26.0	1.415
Standard deviation	9.039	0.1342	9.403	0.138
Variance	81.71	0.0180	88.42	0.019
Skewness	2.04	0.39	1.760	0.395

Sect. 3.2.4). Figure 3.4a shows the experimental variogram of the full set of data (symbols); the estimates lie on a smooth curve. The exchangeable data were sub-sampled to 87 sites, and Fig. 3.4b shows a much more erratic experimental variogram. The result is therefore likely to be less reliable, and it is not clear what kind of curve would fit it best.

Figure 3.4 gives the number of comparisons for each computed semivariance from both the full set of 434 sites and from the sub-sample of 87 sites. The semivariances computed on 434 data, Fig. 3.4a, have more than 1 000 comparisons at the first lag, and they increase to more than 5 000 at the longest lags. The variogram computed from 87 data, Fig. 3.4b, computed with the same step and bin width as in Fig. 3.4a, has many fewer comparisons at all lags. Nevertheless, the number of comparisons at some of the longer lags exceeds 200. Many authors have been misled into thinking that they can obtain reliable estimates of $\gamma(\mathbf{h})$ based on 50 comparisons, or even fewer; they cannot, as is clear from this result with 87 data—variograms computed on small sets of data are unreliable (see Webster and Oliver 1992).

Fig. 3.4 Experimental variograms of the common logarithm of exchangeable potassium, $\log_{10} K^+$, in the topsoil of Broom's Barn Farm, Suffolk; **a** computed from data at 434 sampling quadrats and **b** computed from all 87 quadrats. The *numbers* attached to the *points* are the numbers of paired comparisons from which the semivariances are computed. The *lines* are the best fitting spherical models

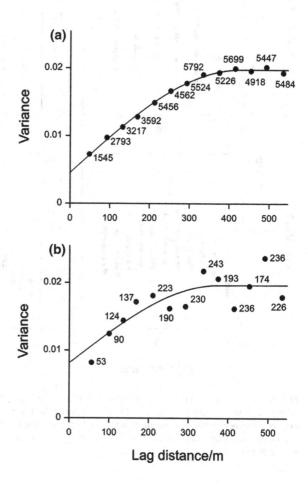

Over the years we have seen many erratic variograms computed on too few data, in some cases as few as 25. Twenty years ago we explored the sampling fluctuation in variograms (Webster and Oliver 1992). We concluded that one should aim for 150 data where variation is isotropic and set 100 as a minimum. Brus and de Gruijter (1994) came to a similar conclusion via a different route, and the message is reinforced with examples in Webster and Lark (2013) and in Oliver and Webster (2014).

For this chapter we have revisited the matter by repeated independent sampling from a much larger correlated random field of $400 \times 400 = 160\ 000$ with an isotropic spherical variogram: $0.283 + 0.700 \times \mathrm{sph}(h|24)$ and variance of 1.0. See Eq. (3.10) for a full definition of the function. Figure 3.5 shows the results of 15

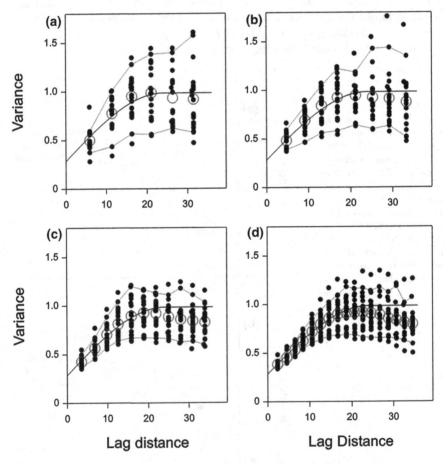

Fig. 3.5 Experimental variograms computed from repeated sampling on grids of 7×7, 9×9, 12×12 and 18×18 points. The *solid lines* are those of the isotropic spherical model fitted to the exhaustive experimental variogram, and the *dashed lines* join the 5 and 95 % quantiles, and the *circles* are the mean values at the lags. The model is $\gamma(h) = 0.283 + 0.700 \times \mathrm{sph}(h|24)$ for a field with variance 1.0

independent repeated samplings for four grids of sizes $7 \times 7 = 49$, $9 \times 9 = 81$, $12 \times 12 = 144$ and $18 \times 18 = 324$ sampling points. Evidently with 144 points the estimates of $\gamma(h)$ at the shorter lag distances lie close to the variogram used to generate the field, the solid curve in the figure, but diverge at the longer lags. The same is true for the much larger samples of 324 points. Sample sizes in the range 100–150 should be adequate where the variogram is required for kriging, but estimates of sill variances will be erratic. Where variation is anisotropic, i.e. not the same in all directions, more data are required to identify it and define it mathematically. We deal with anisotropy below (Sect. 3.2.5).

3.2.2 Sampling Interval and Spatial Scale

The choice of a suitable sampling interval depends on the scale of variation that the practitioner wishes to resolve, e.g. experimental plot, field, farm, catchment, administrative region and so on. If you have rough variograms of the properties of interest or variograms from related ancillary data such as aerial images then choose a sampling interval that will give you at least five estimates of $\gamma(h)$ within the effective range. Alternatively, you can use an accurate existing variogram of a property of interest to determine the kriging errors and so determine an optimal sampling interval for kriging, see Chap. 5 (Burgess et al. 1981; Webster and Lark 2013). If the lag interval exceeds half the range or effective range of variation the resulting variogram is likely to be flat; it will not capture the correlated structure and so will not describe adequately the spatial variation present, as in Fig. 3.6. The experimental variogram of topsoil sand in this figure was computed from a stratified random sample of the soil of the Wyre Forest, England (Oliver and Webster 1987). The average distance between neighbouring sampling points was 165 m, and the experimental variogram was computed with a lag interval of 75 m. The resulting

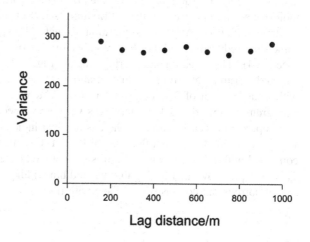

Fig. 3.6 Experimental variogram of topsoil sand from a stratified random survey in the Wyre Forest, England. The variogram is pure nugget

variogram appears as 'pure nugget'—it shows no spatial structure. Further surveys revealed that the range of spatial dependence of topsoil sand here was approximately 70 m. In other words, all of the variation occurs over distances less than 70 m, which is much less than the average sample spacing in the first survey.

3.2.3 Lag Interval and Bin Width

As mentioned above, where data are on a regular grid or at equal intervals on transects the natural step is one interval. Where they are irregularly scattered, the comparisons must be grouped by distance as described in Fig. 3.3. The practitioner must choose both the length of the step, h, and the limits, w, within which the squared differences are averaged for each step. Usually the two are coordinated such that each comparison is placed in one and only one bin. Choosing the width of bins requires judgement. If the steps are short and the bins narrow then there will be many estimates of $\gamma(\mathbf{h})$, which can lead to a 'noisy' variogram because the semi-variances are calculated from few comparisons. If in contrast the steps are large and the bins wide then there might be too few estimates of the semivariances to reveal the form of the variogram. The choice is thus a compromise; it is not one that should be automated. The practitioner should graph the experimental values, as in Fig. 3.7, so that the selection can be made objectively.

We illustrate the effect of lag interval and bin width with irregularly scattered data on cadmium concentrations in the soil of a region to the south east of the Madrid metropolitan area, Spain (Vázquez de la Cueva et al. 2014). The region is 35 km from west to east by 30 km from north to south. The topsoil (0–15 cm) was sampled at 125 sites. The design comprised two superimposed grids, one at 5-km intervals and the other at 1-km intervals. From the possible 1 116 nodes 74 were chosen at random, and 51 points were added 200 m from the 74. At each site five cores of soil were taken from a circle of radius 5 m and bulked for laboratory analysis. Table 3.3 summarizes the statistics. The coefficient of skewness of 1.71 indicates a long upper tail in the distribution (see Sect. 3.4) that might be reduced by transformation. After transformation to natural logarithms the skewness is somewhat reduced to −1.11, but remains outside the generally advised limits (Sect. 3.4). The experimental variograms in Fig. 3.7 were computed from the natural logarithms of cadmium concentrations: those in Fig. 3.7a–c were computed with a lag interval of 1 km and that in Fig. 3.7d with an interval of 3 km. The variogram computed at 1-km intervals is very erratic because of the small number of comparisons in each estimate. There is no clear indication from the experimental values of the kind of model that will fit best. The sequence of points in Fig. 3.7d computed with a lag interval of 3 km is now smoother and has a clearer structure. This example shows that the lag interval and bin width give different pictures of the spatial correlation: contrast Fig. 3.7a–c with d.

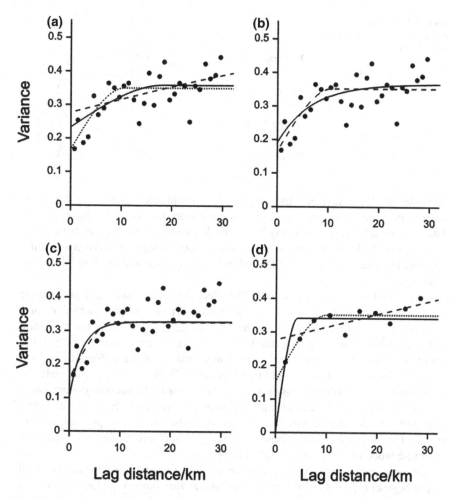

Fig. 3.7 Experimental variograms of cadmium in the topsoil of a region south east of Madrid computed from 125 sampling points, **a–c** at 1-km intervals with bins 1 km wide, and **d** at 3-km intervals with bins 3 km wide. Models have been fitted by GenStat as follows: **a** spherical model to 30 km with range set initially to 25 km (*dashed*), iterated once (*dotted*) and iterated twice (*solid*); **b** spherical model to 30 km with range set initially to 10 km (*dashed*) and exponential model fitted to 30 km with distance parameter set initially to 3 km (*solid*); **c** spherical (*dashed*) and exponential (*solid*) models fitted to 15 km with same initial values for distance parameters as in (**b**); **d** spherical models with range initially set to 3 km (*dashed*) and iterated once (*solid*) and with range set initially to 10 km (*dotted*)

3.2.4 Statistical Distribution

Geostatistical analysis does not require data to follow a normal distribution. However, variograms comprise sequences of variances, and these can be unstable where data are strongly skewed and contain outliers. If your data do not have a

Table 3.3 Summary
statistics of cadmium in soil
of Madrid region

		Cd/mg kg^{-1}	ln(Cd)
Number of data	125		
Minimum		0.005	−5.298
Maximum		0.48	−0.734
Mean		0.137	−2.138
Median		0.11	−2.207
Standard deviation		0.0802	0.589
Variance		0.00643	0.347
Skewness		1.71	−1.107

near-normal distribution and have a skewness coefficient outside the limits ±1, because of a long tail, you should consider transforming them. So, transform the data in some appropriate way, say by taking logarithms, and examine variograms computed on both raw and transformed values. Do the resulting variograms differ substantially apart from a scaling factor? In some cases the answers will be 'no'; in others 'yes'.

Kerry and Oliver (2007a) explored the effects of varying skewness and sample size on simulated random fields with asymmetry. Their results showed that for a large sample size of 1 600 data (on a 5-m grid), the change in shape of the variogram with increasing asymmetry was small, even for a skewness coefficient of 5. For a sample size of 400 (on a 10-m grid), the change in shape of the variograms was not large with increasing skewness and transformation. With 100 data (20-m grid), the semivariances at the first two lags proved to be similar to the generating function of the simulated field, but beyond that they departed progressively as the skewness increased, and for the skewness coefficient of 5 the variogram appeared as pure nugget. Our advice is to transform if it makes a difference to the variogram, but otherwise work with the original data (Table 3.2).

The variogram is sensitive to outliers in the data, i.e. *unexpectedly* large or small values beyond the limits of the main distribution. Box-plots, Fig. 3.8, are an ideal way to identify outliers. All outliers should be investigated and considered as potentially erroneous values before they are allowed to remain as part of the data set. For contaminated sites, however, the largest values will be of most interest. We mentioned above that the same data can contribute to several estimates of $\gamma(\mathbf{h})$, and so outliers inflate the averages. If there are few outliers relative to the whole data, removing them often reduces skewness, and this is a reasonable approach. The values removed can be returned to the data for kriging if desired. Transformation often fails to improve the distribution when outliers are present and can even make matters worse. The alternative is to use one of the robust estimators, such as those of Cressie and Hawkins (1980), Dowd (1984) and Genton (1998).

Cressie and Hawkins's (1980) estimator, $\hat{\gamma}_{CH}(\mathbf{h})$, is based on taking the fourth root of the squared differences and dampens the effect of outliers from the secondary process. It is given by

Fig. 3.8 Box-plot computed from a field of 400 values simulated with a spherical variogram function with zero nugget and contaminated with five outliers resulting in a skewness coefficient of 1.5, where *filled squares* represent the far outliers which are three times beyond the interquartile range and *filled circles* are near outliers

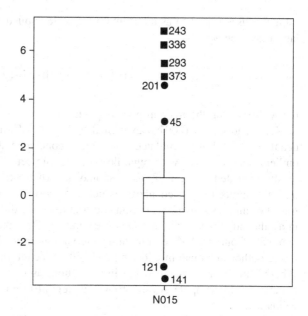

The denominator in Eq. (3.3) is a correction based on the assumption that the underlying process to be estimated has normally distributed differences over all lags.

$$2\hat{\gamma}_{CH}(\mathbf{h}) = \frac{\left\{\frac{1}{m(\mathbf{h})}\sum_{i=1}^{m(\mathbf{h})}|z(\mathbf{x}_i) - z(\mathbf{x}_i + \mathbf{h}|^{\frac{1}{2}}\right\}^4}{0.457 + \frac{0.494}{m(\mathbf{h})} + \frac{0.045}{m^2(\mathbf{h})}}. \tag{3.3}$$

Dowd's (1984) estimator, $\hat{\gamma}_D(\mathbf{h})$, and Genton's, $\hat{\gamma}_G(\mathbf{h})$, estimate the variogram for a dominant intrinsic process in the presence of outliers. Dowd's estimator is given as

$$2\hat{\gamma}_D(\mathbf{h}) = 2.198\{\text{median}(|y_i(\mathbf{h})|)\}^2, \tag{3.4}$$

where $y_i(\mathbf{h}) = z(\mathbf{x}_i) - z(\mathbf{x}_i + \mathbf{h})$, $i = 1, 2,..., m(\mathbf{h})$. The term within the braces of Eq. (3.4) is the median absolute pair difference (MAPD) for lag $\mathbf{h}$, which is a scale estimator only for variables where the expectation of the differences is zero. The constant is a correction that scales the MAPD to the standard deviation of a normally distributed population.

Genton's (1998) estimator, $\hat{\gamma}_G(\mathbf{h})$, is based on the scale estimator, Q_{Nh}, of Rousseeuw and Croux (1992). The estimator, Q_{Nh}, is given by

$$Q_{Nh} = 2.219\{|X_i - X_j|; i < j\}_{\binom{H}{2}}, \tag{3.5}$$

where the constant 2.219 is a correction for consistency with the standard deviation of the normal distribution, and H is the integral part of $(N/2) + 1$. Genton's (1998) estimator uses Eq. (3.5) as an estimator of scale applied to the differences at each lag; it is given by

$$2\hat{\gamma}_G(\mathbf{h}) = \left[2.219\{|y_i(\mathbf{h}) - y_j(\mathbf{h})|; i < j\}_{\left(\frac{H}{2}\right)}\right]^2, \qquad (3.6)$$

but with H being the integral part of $\{m(\mathbf{h}/2)\} + 1$.

Kerry and Oliver (2007b) examined the effects of outliers and sample size in detail with fields of simulated data. They concluded that skewness caused by outliers must be dealt with regardless of the number of data. Furthermore, their results indicated that practitioners should act when skewness exceeds 0.5 rather than the limits mentioned above which are those generally used. Although the robust estimators provided a reasonable solution, they did not perform equally well in all the situations Kerry and Oliver examined. They therefore recommended the removal of outliers before computing the variogram as the current 'best practice' where outliers are randomly located and will not be returned to the data for kriging. Where outliers are crucial to the investigation, as on contaminated sites, practitioners should compute several robust variograms and compare them by cross-validation.

A field of 400 values was simulated on a 10-m grid by a spherical function with zero nugget, a sill of 1 and range of 75 m (Kerry and Oliver 2007b), i.e. $c_0 = 0$, $c = 1$ and $r = 75$ m, see Eq. (3.10). Five of the values were contaminated by another process to give a skewness coefficient of 1.5. Figure 3.8 shows the box-plot of these values; the outliers are >4. An experimental variogram was computed from all the values and modelled, Fig. 3.9a. The nugget variance has increased dramatically to 0.617 showing the effect of adjacent disparate values. The sill variance is 1.341, which is an expression of the increase in variance, and the range has decreased to 67.5 m. The dashed line in Fig. 3.9a is the generating function of the simulated field. Figure 3.9b shows the experimental variogram and model for the same values, but with the outliers removed. The nugget variance is zero, the sill variance is almost 1.0 and the range is 73.6 m. This result shows how important it is to deal with outliers in data.

3.2.5 Anisotropy

Variation can vary from one direction to another, i.e. it can be anisotropic. You should therefore check your data for fluctuations in directional variation. In many instances the anisotropy is such that it could be made isotropic by a simple linear transformation of the spatial coordinates. Imagine that the region sampled is placed on a rubber sheet, which could be stretched in the direction in which variation seemed shortest. If the stretching eventually produces variation that is the same in that direction as that in the perpendicular direction then the anisotropy is known as *geometric*. The equation for the transformation is

Fig. 3.9 Experimental
variograms (*symbols*) and
fitted models (*solid lines*)
computed from a field of 400
values simulated with a
spherical variogram function
with zero nugget (*dashed
line*): **a** contaminated with
five outliers resulting in a
skewness coefficient of 1.5
and **b** with the outliers
removed

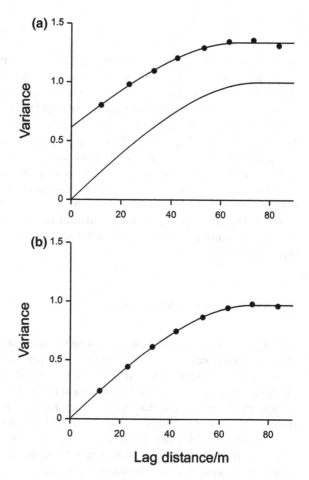

$$\Omega(\vartheta) = \left\{ A^2\cos^2(\vartheta - \varphi) + B^2\sin^2(\vartheta - \varphi) \right\}^{1/2}, \tag{3.7}$$

where Ω defines the anisotropy, φ is the direction of maximum continuity and ϑ is
the direction of the lag.

For a spherical or exponential variogram, A is the distance parameter in the
direction of greatest continuity, i.e. the maximum value, and B is the distance
parameter in the direction of least continuity or greatest variation, the minimum. For
an unbounded variogram, the roles of A and B are reversed, and A has the larger
gradient in the direction of the greatest rate of change and B has the smaller gradient
in the direction of least change. Figure 3.12 shows an example in which there are
differences in the ranges for a bounded variogram (see Sect. 5.1).

Anisotropy can also occur as preferentially orientated zones with different means
that result in changes in variance with change in direction and fluctuations in the
sill. This is known as *zonal anisotropy*.

3.2.6 Trend

In Chap. 1 we mentioned trend. We consider it briefly here in relation to the variogram, but Chap. 6 is devoted to the matter, and readers should turn to that chapter for detail. We can always calculate an experimental variogram by Eq. (3.1), but it estimates the theoretical variogram $\gamma(\mathbf{h})$ only where the underlying process is random. If there is trend then this equation gives a false summary of the random part of the process. Typically, where trend is present the experimental variogram increases without bound, and if it dominates then the experimental sequence becomes increasingly steep as the lag distance increases (see Fig. 6.11). If you obtain such a result then examine your data by fitting simple linear and quadratic polynomials on the coordinates. Alternatively, map the data by some simple graphical procedure before doing a statistical analysis; if the map shows gradual continuous change across the region then there is trend with more or less patchiness superimposed.

3.3 Modelling the Variogram

The experimental variogram consists of semivariances at a finite set of discrete lags. These semivariances are estimates based on samples; they are therefore subject to error, which itself varies from one estimate to the next. In addition, the underlying function is continuous for all $\mathbf{h}$, Eqs. (3.5) and (3.8). The next step in variography is to fit a smooth curve or surface to the experimental values, one that describes the principal features of the sequence (see Sect. 3.3.1) while ignoring the point-to-point erratic fluctuation. Not any plausible-looking curve or surface will serve; it must have a mathematical expression that can legitimately describe the variances of random processes. It must guarantee non-negative variances of combinations of values, and there are only a few simple functions that do so. They are known as *conditional negative semi-definite* (CNSD) because the matrices to which they contribute are themselves conditional negative semi-definite (see Webster and Oliver 2007, for a full account).

3.3.1 Principal Features of the Variogram

(1) An increase in variance with increasing lag distance from the ordinate
     In Fig. 3.10a the variogram shows a monotonic increase in variance as the lag distance increases. The slope shows the change in the *spatial autocorrelation* or *dependence* between sampling points as the separation distance increases. In other words at short lag intervals, $|\mathbf{h}|$, the semivariances, $\gamma(|\mathbf{h}|)$, are small indicating that values of $Z(\mathbf{x})$ are similar, and as $|\mathbf{h}|$ increases they become increasingly dissimilar on average.

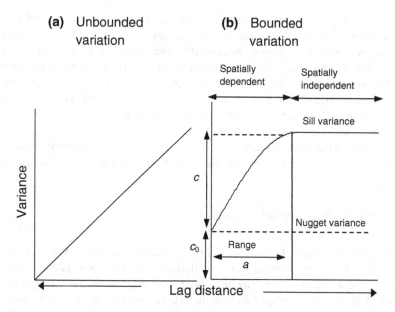

Fig. 3.10 Examples of: **a** unbounded and **b** bounded variogram models with annotations to illustrate the parameters of a bounded model function

(2)   An upper bound, the *sill variance*

   If the process is second-order stationary then the variogram will reach an upper bound, the *sill variance*, after the initial increase as in Fig. 3.10b. For some variograms the sill remains constant, whereas for others it is an asymptote, which we explain below. The sill variance is also the *a priori* variance, σ^2, of the process.

(3)   The *range* of spatial correlation or dependence

   A variogram that reaches its sill at a finite lag distance has a *range*, which is the limit of spatial correlation where the autocorrelation becomes 0, Fig. 3.10a. Places further apart than this are spatially uncorrelated or independent. Variograms that approach their sills asymptotically have no strict ranges; in practice, however, we use an effective range at the lag distances where they reach 0.95 of their sills.

(4)   *Unbounded* variogram

   The variogram may increase indefinitely with increasing lag distance as in Fig. 3.10a. It describes a process that is not second-order stationary, and the covariance does not exist. The variogram, however, does exist and fulfils Matheron's (1965) intrinsic hypothesis (see Sect. 2.1, Chap. 2).

(5)   A positive intercept on the ordinate, the *nugget variance*

   The variogram often approaches the ordinate with a positive intercept known as the *nugget variance*, Fig. 3.10b. Theoretically, when $\mathbf{h} = 0$ the semivariance should also be 0 (see Chap. 2). The term 'nugget' in this context was coined in gold mining because gold nuggets appear to occur at random and independently of one another. They represent a discontinuity in the variation, an uncorrelated component, because the gold content no longer relates to that at neighbouring sites. For properties that vary continuously in space, such as the amount of water vapour in the atmosphere or the pH of the soil, the nugget variance arises from measurement error (usually a small component) and variation over distances less than the shortest sampling interval.

(6) Directional variation *anisotropy*

Spatial variation might vary according to direction, as mentioned above, and we need to be able to take this into account in our analysis and modelling.

3.3.2 Variogram Model Functions

There are two principal kinds of function, namely bounded and unbounded (Fig. 3.10). We give the equations and illustrate the three most popular models; power function (unbounded), spherical (bounded) and exponential (asymptotically bounded). If none of these appears to fit the experimental values then more complex functions may be fitted. Such functions may be any combination of simple CNSD functions; these combinations are themselves CNSD.

Theoretically, the variogram model should intercept the ordinate at the origin according to theory as in Eq. (3.8) and Fig. 3.10a. In practice the experimental variogram frequently, indeed usually, appears to approach the ordinate at some positive finite value. To make the curve fit one adds a nugget component to the simple function as in Eqs. (3.8)–(3.11) and Fig. 3.10b where there is a nugget variance and a structured component. A more complex function is required where there are two or more distinct scales of spatial dependence, i.e. a nested model. We illustrate this scenario with two spherical functions, one nested within the other, plus a nugget variance, Eq. (3.12) and Fig. 3.11. We describe the models in their isotropic form; they are symmetric about zero lag, but we define them for $|\mathbf{h}| \geq 0$ only.

The equations for the four models are as follows.

Power function. This is an unbounded function

$$\gamma(h) = gh^{\beta} \qquad \text{for } 0 < \beta < 2, \qquad (3.8)$$

where g describes the intensity of the variation and β describes the curvature. If $\beta = 1$, the variogram is linear and g represents the gradient. The limits 0 and 2 are excluded because $\beta = 0$ indicates constant variance for all $h > 0$ and $\beta = 2$ that the function is parabolic with zero gradient at the origin. The latter means that the process is not random. Figure 3.10a gives an example of an unbounded variogram with no nugget variance as in the equation above. If such a function had a positive intercept at the ordinate the equation would be

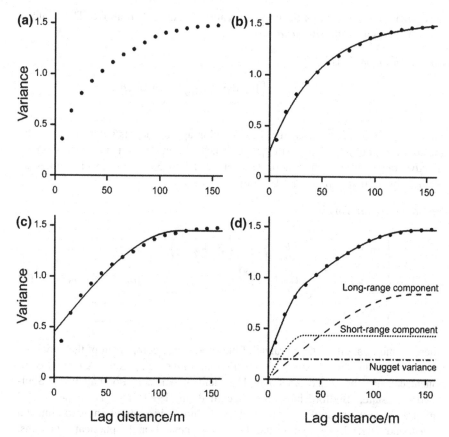

Fig. 3.11 Wheat yield recorded in 1999 in Football Field on the Shuttleworth Estate **a** the experimental variogram (*symbols*), **b** the *solid line* is an exponential function fitted to the experimental values, **c** the *solid line* is a spherical function fitted to the experimental values and **d** the best fitting nested spherical model (*solid line*). The model was decomposed to illustrate the individual model components as shown by the *ornamented lines*

$$\gamma(h) = c_0 + gh^\beta \qquad \text{for } 0 < \beta < 2, \tag{3.9}$$

Spherical model. This is

$$\gamma(h) = \begin{cases} c_0 + c\left\{\frac{3h}{2r} - \frac{1}{2}\left(\frac{h}{r}\right)^3\right\} & \text{for } 0 < h \leq r \\ c_0 + c & \text{for } h > r \\ 0 & \text{for } h = 0, \end{cases} \tag{3.10}$$

where c_0 is the nugget variance, c is the variance of spatially correlated component and r is the range of spatial dependence. Figure 3.10b illustrates a spherical variogram with annotations of the main features as described above. The quantity $c_0 + c$ is known as the *sill variance*.

Exponential model. This is

$$\gamma(h) = \begin{cases} c_0 + c\{1 - \exp\left(-\frac{h}{a}\right)\}, & \text{for } 0 < h \\ 0 & \text{for } h = 0 \end{cases} \qquad (3.11)$$

where a is the distance parameter. This function approaches its sill asymptotically, and so it does not have a finite range. For practical purposes it is usual to assign an effective range, a', which is approximately equal to $3a$. Figure 3.11b shows an example of a fitted exponential function.

Nested spherical. This is

$$\gamma(h) = \begin{cases} c_0 + c_1\left\{\frac{3h}{2r_1} - \frac{1}{2}\left(\frac{h}{r_1}\right)^3\right\} + c_2\left\{\frac{3h}{2r_2} - \frac{1}{2}\left(\frac{h}{r_2}\right)^3\right\} & \text{for } 0 < h \le r_1 \\ c_0 + c_1 + c_2\left\{\frac{3h}{2r_2} - \frac{1}{2}\left(\frac{h}{r_2}\right)^3\right\} & \text{for } r_1 < h \le r_2 \quad (3.12) \\ c_0 + c_1 + c_2 & \text{for } h > r_2 \\ 0 & \text{for } h = 0, \end{cases}$$

where c_1 and r_1 are the sill and range of the short-range component of the variation, and c_2 and r_2 are the sill and range of the long-range component. A nugget component can also be added as above. The yield of wheat in Football Field, Shuttleworth Estate, Bedfordshire, was recorded in 1999, and an experimental variogram was computed from the values. Figure 3.11a shows the experimental values and Fig. 3.11b and c the fitted exponential and spherical functions, respectively. It is clear that the spherical function, Fig. 3.11c, fits poorly and that the exponential model, Fig. 3.11b, fits reasonably. The fit of the latter emphasizes the small change in slope evident in the experimental variogram at about lag 30 m and another change at around lag 140 m. Figure 3.11d shows the nested spherical function, which provides a near-perfect fit to the experimental values with a smaller nugget variance than the exponential model and follows the values closely. The variously ornamented lines in Fig. 3.11d show the components of the nested model; the nugget, short-range and long-range. Table 3.4 gives the parameters of these models; they show that the spherical function has a larger nugget variance than the other two models and a smaller range of spatial dependence. The parameters of the exponential model are closer to those of the nested spherical with a smaller nugget variance and an approximate effective range ($3a$) of 140 m. The diagnostics in Table 3.4 reflect the visual observations. The residual sum of squares (RSS) is much larger for the spherical function than for the exponential and nested spherical models, and that for the exponential is larger than for the nested model.

If your models have the same number of parameters and the ones fitted seem to fit well then choose the one with the smallest residual sum of squares (RSS) or smallest mean square. You may wish to fit more complex models, but you should

Table 3.4 Parameters of models fitted to yield from Football Field, Shuttleworth Estate, Bedfordshire, UK recorded in 1999

Model type	Estimates of parameters						Diagnostics			
	c_0	c_1	c_2	a_1/m	a_2/m	r/m	RSS	RMS	% variance	AIC
Exponential	0.2437	1.291				47.33	3 802	292.5	99.5	96.86
Spherical	0.4516	0.9994		118.4			14 896	1 146	97.8	118.7
Nested spherical	0.1975	0.4318	0.8415	33.88	137.8		1 052	95.63	99.8	82.97

be cautious because you can always diminish the RSS by increasing the number of parameters in the fitted model. For example, the double spherical model with nugget has five parameters, whereas the simpler single spherical model with nugget has only three. Are the two additional parameters justifiable? To ensure parsimony in our fitting we can compute an estimate of the Akaike Information Criterion (AIC) (see Webster and Oliver 2007, for more detail) if, as in our comparisons above, the models have unequal numbers of parameters as for the nested spherical model. The AIC is estimated by

$$\text{AIC} = \left\{ n \ln \left(\frac{2\pi}{n} \right) + n + 2 \right\} + n \ln R + 2p, \qquad (3.13)$$

where n is the number of points on the variogram (16 in this example), p is the number of model parameters and R is the mean square of the residuals (RMS in Table 3.4). The quantity in braces is constant for any one experimental variogram, and so we need compute only

$$\hat{A} = n \ln R + 2p. \qquad (3.14)$$

We then choose the model for which $\hat{A}$ is the least. In Table 3.4, $\hat{A}$ is markedly smaller for the nested spherical model than for the exponential and spherical functions, and so we would choose the more complex function as providing the best fit.

Anisotropic model
To examine data for both types of anisotropy compute the variogram in at least four directions to start with: along the rows, down the columns and on the principal diagonals if data are on a rectangular grid (see Fig. 3.12). The semivariances can be plotted in these directions, and no information is lost. For irregularly scattered data, we have to group the separations by direction as well as distance as in Fig. 3.3. The angle, α, within which data are included in estimating the semivariance should allow complete cover to start with, i.e. $\pi/4$ for four angles, which will include all data in those directions. Note, however, that this procedure loses some directional information. If it reveals directional variation then reduce α to identify the direction of strongest anisotropy, but realize that the smaller α becomes the fewer will be the

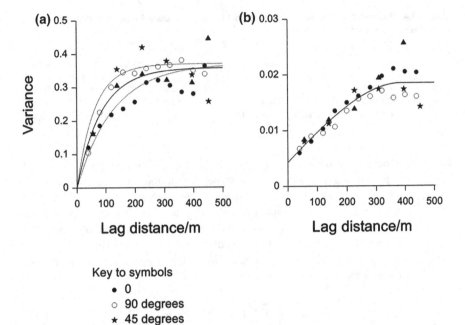

Fig. 3.12 Experimental variograms computed in four directions: **a** pH; the *solid line* is the isotropic exponential model and the *dotted lines* form the envelope of the fitted anisotropic exponential model and **b** $\log_{10} K^+$ with the fitted isotropic spherical function

number of comparisons and the greater will be the error in the estimated semi-variances. Choosing α is therefore a compromise between a stable estimate based on many comparisons that will underestimate the directional effect with a wide angle and one that is subject to greater error but reflects the anisotropy more closely.

Figure 3.12a, b shows the experimental variograms of pH and exchangeable potassium (as $\log_{10} K^+$), respectively, at Broom's Barn Farm computed in four directions. The directional variogram of pH shows a longer range of variation in the north–south (90°) direction and a shorter range in the east–west (0) direction, whereas for $\log_{10} K^+$ no anisotropy is evident. The directional variogram for pH has been fitted with an anisotropic exponential function:

$$\gamma(h, \vartheta) = c_0 + c\{1 - \exp[-|\mathbf{h}|/\Omega(\vartheta)]\}, \tag{3.15}$$

where $|\mathbf{h}|$ is the modulus of the lag and $\Omega(\vartheta)$ is defined in Eq. (3.7). The model parameters are given in Table 3.4 and Fig. 3.12a shows the envelope of the model as the dotted lines. An isotropic exponential function was also fitted; the parameters of this are given in the table and the model is the solid line in Fig. 3.12a.

3.4 Factors Affecting the Reliability of Variogram Models

There are operational aspects that we need to consider when computing the experimental variogram and fitting models. They include the effects of poor choice of lag or bin interval and of maximum lag, and sample size on the reliability of the model parameters that will then be used for kriging. The experimental variogram should be computed and modelled only as far as it is reliably estimated. We recommend that you compute it to a maximum lag of no more than a third to one half of the extent of the data. Table 3.1 shows how the number of comparisons (counts) starts to decrease after a certain lag distance. It is at this lag distance (about 530 m) that the semivariances also start to depart from the smooth curve; this is a sign that the estimates are becoming increasingly unreliable (Fig. 3.13). Table 3.5 shows how the model parameters of the fitted spherical models also change for $\log_{10} K^+$ when the model was fitted to a maximum lag of 900 m compared with 550 m.

3.4.1 Fitting Models

Fitting models remains controversial in geostatistics, yet it is one of the most important stages to get right. Some practitioners fit models by eye, which we do not recommend because the observed semivariances may fluctuate too much from point to point and their accuracy is not constant, which makes this approach unreliable. Fitting models with 'black box' software can also produce poor results because there is no choice, judgement or control over the process. We recommend a procedure that involves both visual inspection and statistical fitting in steps as follows.

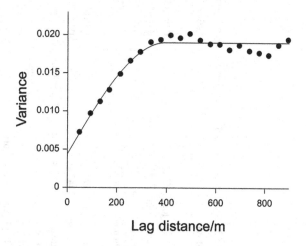

Fig. 3.13 Experimental variogram computed and modelled to a maximum lag of 900 m for $\log_{10} K^+$ at Broom's Barn Farm

Table 3.5 Parameters of models fitted to exchangeable potassium in the topsoil at Broom's Barn Farm, England

Model type	Estimates of parameters				Diagnostics		
	c_0	c	a/m	r/m	RSS	RMS	% variance
Spherical—$\log_{10}$ K$^+$87	0.008113	0.01151	376.4		0.01184	0.001184	40.9
Spherical—$\log_{10}$ K$^+$434	0.004535	0.01524	426.0		0.003683	0.000368	99.4
Exponential—$\log_{10}$ K$^+$87	0.001351	0.01860		110.4	0.01127	0.001127	43.8
Exponential—$\log_{10}$ K$^+$434	0.001578	0.01981		180.0	0.01943	0.001943	96.7
	w	α					
Power—$\log_{10}$ K$^+$87	0.01092	0.2484			0.01254	0.001140	43.1
Power—$\log_{10}$ K$^+$443	0.00778	0.3876			0.05727	0.005206	91.2
Spherical—$\log_{10}$ K$^+$434 Maximum lag 900 m	0.004399	0.01458	393.2				96.4
Exponential—pH	0	0.3588		85.2	13.67	0.5256	78.7

Model type	Estimates of parameters				Diagnostics		
	c_0	c	φ (rad)		A/m	B/m	A/B
Anisotropic exponential—pH	0	0.371	2.735 (156.7°)		126.9	61.96	2.048
					RSS	RMS	% variance
					9.255	0.3856	84.4

1. First, plot the experimental variogram, the black discs in Fig. 3.14.
2. Choose several models with a similar shape and fit each in turn by weighted least squares, the curves in Fig. 3.14.
3. Plot the fitted models on the graph of the experimental variogram and assess whether the fit looks reasonable. If all plausible models seem to fit well, choose the one with the smallest residual sum of squares (RSS) or smallest mean square. If the models have unequal numbers of parameters as for the nested spherical model then compute the Akaike Information Criterion (AIC) and choose the model for which the AIC is least as above.

Figure 3.14a, c and e shows the experimental variogram computed from $\log_{10} K^+$ with 87 data. None of the three models chosen, power, spherical and exponential Eqs. (3.9)–(3.11), respectively, and displayed above in Sect. 3.3.2, appears to fit well. Without the diagnostic information in Table 3.5 it would be difficult to choose between them. The exponential function has the smallest residual mean square (RMS) and accounts for the most variance, albeit only 44 %. The difference between the parameters of the two bounded functions, spherical and exponential, Eqs. (3.10) and (3.11), is marked, especially in relation to the nugget variance, c_0. The power function, Eq. (3.7), provides the next best fitting model, although it is clear from the variogram of the full set of data, Fig. 3.14b, that the underlying process is second-order stationary and requires a bounded function. For the same functions fitted to the experimental variogram of the full data, 434 sites, the best fitting function is clearly the spherical one which has a very small RMS and accounts for 99.4 % of the variance (Table 3.5). The exponential and power functions fit less well both visually and from the diagnostic values. The importance of an adequate sample size is clear from this example, which illustrates the poor fit of all functions to the experimental values from the sample of 87 and the small percentage variance accounted for compared with those for the full set.

Finally, we compare the effect of choice of lag interval and bin width for the data on the cadmium in the soil near Madrid, again with data from Vázquez de la Cueva et al. (2014). Figure 3.7 shows experimental variograms computed with a lag interval of 1 km in Fig. 3.7a–c and of 3 km in Fig. 3.7d. The variogram computed with a lag of 1 km is so erratic that none of the functions provides a good fit. Several of the exponential and spherical models fitted appear to be as good as any other, whereas for Fig. 3.7d it is clear that the model represented by the dotted line (spherical with range of 10 km and no iteration) provides the best fit. Table 3.6 lists the parameters of the functions fitted to the two experimental variograms. Different initial values for the non-linear parameter, r, for the spherical model were used and also different numbers of iterations which give increasing weight to values near to the origin. Because the experimental semivariances are based on different numbers of paired comparisons, $m(\mathbf{h})$ in Eq. (3.1), and because confidence in the estimate of variance decreases as its value increases, we generally weight the semivariances by the number of counts when fitting the models. The inverse relation between the reliability of an estimate of variance and the variance itself led Cressie (1985) to propose a more elaborate weight, which has the form

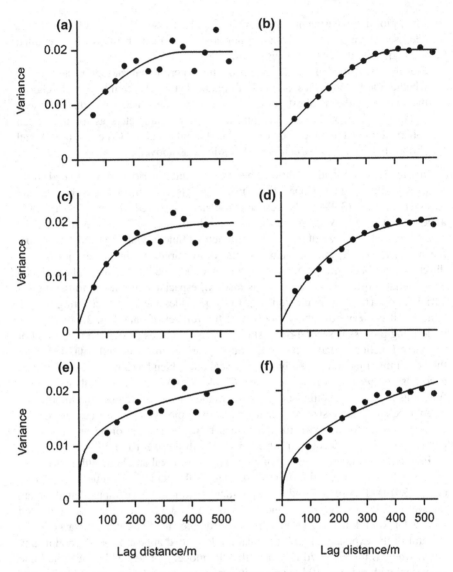

Fig. 3.14 Experimental variograms computed from 87 data for $\log_{10}$ K$^+$ Broom's Barn Farm, Suffolk and fitted with: **a** spherical model, **c** exponential model and **e** power function, and experimental variograms computed from 434 data and fitted with: **b** spherical model, **d** exponential model and **f** power function

$$m(\mathbf{h}_j)\big/\gamma^{*2}(\mathbf{h}_j),\qquad\qquad(3.16)$$

where $\gamma^{*2}(\mathbf{h}_j)$ is the value of semivariance predicted by the model. The quantity $\gamma^{*2}(\mathbf{h}_j)$ is inserted into the weighting vector and the fitting is repeated, and the

Table 3.6 Models fitted to experimental variograms of cadmium in soil of Madrid region

Fitting	Estimates of parameters				Diagnostics		
	c_0	c	a/km	r/km	MSE	MSDR	Correlation
Lag interval 1 km							
Spherical to 30 km, initial 25 km	0.2766	0.2011	76.0		0.3422	1.112	0.144
After one iteration	0.2329	0.1259	18.5		0.3480	1.201	0.145
After two iterations	0.1665	0.1833	10.6		0.3581	1.341	0.167
Spherical to 30 km, initial 10 km	0.1654	0.1868	10.75		0.3594	1.350	0.164
Exponential to 30 km, initial 3 km	0.1907	0.1762		6.63	0.3424	1.219	0.1898
Spherical to 15 km, initial 10 km	0.1575	0.1685	8.57		0.3392	1.284	0.228
Exponential to 15 km, initial 3 km	0.1012	0.2261		2.54	0.3371	1.292	0.241
Lag interval 3 km							
Spherical to 30 km, initial 3 km	0.2754	0.2062	76.8		0.3422	1.115	0.144
After iteration	0	0.3415	4.32		0.3602	2.125	0.258
Spherical to 30 km	0.1536	0.1978	10.08		0.3574	1.364	0.181

c_0 is the nugget variance, c is the sill variance of the correlated structure and a is the range of the spherical model and r is the distance parameter of the exponential model
MSE is the mean squared error and MSDR is the mean squared deviation ratio

whole process is iterated to convergence, i.e. until there is no perceptible change in $\gamma^{*2}(\mathbf{h}_j)$. However, McBratney and Webster (1986) discovered that one iteration was usually sufficient, and so in GenStat, for example, only one repeat is programmed.

The second iteration is a refinement of the former proposed by McBratney and Webster (1986):

$$m(\mathbf{h}_j)\hat{\gamma}(\mathbf{h}_j)/\gamma^{*3}(\mathbf{h}_j), \qquad (3.17)$$

where $\hat{\gamma}(\mathbf{h}_j)$ is the observed value of the semivariance at $\mathbf{h}_j$. Both iterations give more weight to estimates close to the origin, which is usually desirable for kriging.

The initial value of the non-linear parameter, r, for the spherical model, can seriously affect the final model fitted: contrast the three curves in Fig. 3.14d. The weights given to the semivariances, $\hat{\gamma}(h)$, can also seriously affect the final model if the distance parameter is chosen poorly initially: see Fig. 3.7a and d and the parameters in Table 3.6.

Another fairly popular way of choosing models for variograms is by cross-validation. This procedure involves leaving out each and every value in the data in turn and kriging the value there using the surrounding data and the given model parameters. The kriged values $\hat{Z}(\mathbf{x}_i)$ are compared with the observed ones $z(\mathbf{x}_i)$. The mean squared error (MSE) between the predictions and the observed values, the mean squared deviation ratio (MSDR) and the median of the squared deviation ratio are calculated and used as criteria of the goodness of the models. The precise nature of these quantities will be apparent when we have described kriging and so they are defined at the end of the next chapter, Chap. 4.

Chapter 4
Geostatistical Prediction: Kriging

Abstract Kriging is the geostatistical method of prediction. It is a best linear unbiased predictor on punctual or block supports; best in the sense that its prediction error variances are minimized. It is in practice a weighted moving average in which the weights depend on the variogram and the configuration of the sample points within the neighbourhood of its targets. Ordinary kriging is by far the most popular method, partly because it is robust with respect to departures from the underlying assumptions. There are, however, numerous more advanced types of kriging for specific tasks. Examples illustrate the effects on the kriging weights, the predictions and the prediction variances of changing the variogram parameters and the sample configurations. Punctual and block kriging are compared for mapping the predictions and errors. Kriging in the presence of anisotropy, simple kriging and lognormal kriging are also illustrated. A solution for back-transformation to the original scale is given for lognormal kriging. Punctual kriging can be used to identify suitable variogram models from the diagnostic statistics of 'leave-one-out' cross-validation.

Keywords Interpolation · Best linear unbiased predictor · Ordinary kriging · Lognormal kriging · Simple kriging · Punctual kriging · Block kriging · Weights · Neighbourhood · Mapping · Cross-validation

4.1 Introduction

In Chap. 1 we made the point that the environment is continuous and that scientists can measure its properties at only finite numbers of places on small supports. Yet those scientists or their clients often want to know values of those properties at unvisited places in between, in principle everywhere, and to map them; they want to interpolate from their measurements.

Several mathematical interpolators and regression (trend surface analysis) have been used with varying success for making maps from sparse data. None, however, provides sound estimates of the errors in its interpolations. Kriging, the geostatistical method of interpolation, does that. Further, it minimizes the errors and is best in that sense, and because its predictions are also unbiased it is often known as a best linear unbiased predictor (BLUP).

The term *krigeage* was coined by P. Carlier in recognition of D.G. Krige's pioneering innovation for estimating concentrations of gold and other metals in ore bodies. Matheron (1963) later introduced it into the English language as 'kriging', and his doctoral thesis (Matheron 1965) placed the technique within the general framework of the theory of random processes. Matheron's work was not in isolation; Kolmogorov (1939, 1941), Wold (1938) and Wiener (1949) had already come close to kriging, but in time rather than in space (Cressie 1990).

Kriging predicts values at unvisited sites from sparse sample data based on a stochastic model of continuous spatial variation. It does so by taking into account knowledge of the spatial variation as represented in the variogram or covariance function. Ordinary kriging requires no other information than that plus the measurements and their geographic coordinates. It is by far the most popular kind of kriging, and with good reason; it serves well in most situations with its assumptions easily satisfied. It is also robust with regard to moderate departures from those assumptions, and therefore we focus on it. More elaborate forms of kriging have been developed to tackle increasingly complex problems in petroleum engineering, mining and geology, meteorology, soil science, precision agriculture, pollution control, public health, fishery, plant and animal ecology, remote sensing and hydrology. We devote another chapter, Chap. 6, to kriging with drift, but others are beyond the scope of this brief, and we refer you to our fuller text (Webster and Oliver 2007) for descriptions of several of them.

4.2 Theory

Ordinary kriging is based on the assumption that variation is random and spatially dependent, and that the underlying random process is intrinsically stationary with constant mean and a variance that depends only on separation in distance and direction between places and not on absolute position. The assumption is the same as that on which the variogram is based—see Eqs. (2.1) and (2.4).

A kriged prediction is a linear sum of data, which may be in one, two or three dimensions. Most applications in environmental science are in two dimensions, and so we treat the matter as two-dimensional. Predictions can be made for points (i.e. having the same support as the measurements) or blocks, and we formalize the procedure for both punctual and block kriging below.

Suppose that values of a random variable, Z, have been recorded at sampling points, $\mathbf{x}_1, \mathbf{x}_2, \ldots, \mathbf{x}_N$ to give N data, $z(\mathbf{x}_i)$, $i = 1, 2, \ldots, N$. For punctual kriging, we predict Z at any new point, $\mathbf{x}_0$, by

$$\hat{Z}(\mathbf{x}_0) = \sum_{i=1}^{N} \lambda_i z(\mathbf{x}_i), \tag{4.1}$$

where λ_i are the weights. To ensure that the estimate is unbiased the weights are made to sum to 1:

$$\sum_{i=1}^{N} \lambda_i = 1. \tag{4.2}$$

The expected difference is $\mathrm{E}[\hat{Z}(\mathbf{x}_0) - z(\mathbf{x}_0)] = 0$, and the prediction variance is given by

$$\mathrm{var}\left[\hat{Z}(\mathbf{x}_0)\right] = \mathrm{E}\left[\{\hat{Z}(\mathbf{x}_0) - z(\mathbf{x}_0)\}^2\right]$$
$$= 2\sum_{i=1}^{N} \lambda_i \gamma(\mathbf{x}_i - \mathbf{x}_0) - \sum_{i=1}^{N}\sum_{j=1}^{N} \lambda_i \lambda_j \gamma(\mathbf{x}_i - \mathbf{x}_j), \tag{4.3}$$

where the quantity $\gamma(\mathbf{x}_i - \mathbf{x}_0)$ is the semivariance of Z between the sampling point $\mathbf{x}_i$ and the target point $\mathbf{x}_0$ and $\gamma(\mathbf{x}_i - \mathbf{x}_j)$ is the semivariance between the ith and jth sampling points. The semivariances are derived from the variogram model, partly because there is no measure of the semivariances between the data points and the target points where we have no observed values and partly because only by doing so can we guarantee that the variances are not negative. If a target point also happens to be a sampling point then punctual kriging returns the observed value there and the estimation variance is zero. Punctual kriging is an exact interpolator in this sense (see Sect. 4.2.5).

Practitioners often want to predict average values within areas that are larger than the supports of the data; for that they need block kriging. The above formulae for ordinary punctual kriging are readily adapted to block kriging. The estimate for any block is still a weighted average of the data, $z(\mathbf{x}_1)$, $z(\mathbf{x}_2)$, ..., $z(\mathbf{x}_N)$:

$$\hat{Z}(B) = \sum_{i=1}^{N} \lambda_i z(\mathbf{x}_i), \tag{4.4}$$

and the weights sum to 1 as before. The prediction variance of $Z(B)$, however, is now

$$\mathrm{var}\left[\hat{Z}(B)\right] = \mathrm{E}\left[\{\hat{Z}(B) - z(B)\}^2\right]$$
$$= 2\sum_{i=1}^{N} \lambda_i \bar{\gamma}(\mathbf{x}_i, B) - \sum_{i=1}^{N}\sum_{j=1}^{N} \lambda_i \lambda_j \gamma(\mathbf{x}_i - \mathbf{x}_j) - \bar{\gamma}(B, B), \tag{4.5}$$

where $\bar{\gamma}(\mathbf{x}_i, B)$ is the average semivariance between data point $\mathbf{x}_i$ and the target block B, and $\bar{\gamma}(B, B)$ is the average semivariance within B, the within block variance.

The next step in kriging is to find the weights that minimize the kriging variances subject to the constraint that they sum to 1. Equations (4.1)–(4.3) for a point lead to a set of $N + 1$ equations in the $N + 1$ unknowns:

$$\sum_{i=1}^{N} \lambda_i \gamma(\mathbf{x}_i - \mathbf{x}_j) + \psi(\mathbf{x}_0) = \gamma(\mathbf{x}_j - \mathbf{x}_0) \quad \text{for all } j$$

$$\sum_{i=1}^{N} \lambda_i = 1. \tag{4.6}$$

The quantity $\psi(\mathbf{x}_0)$ is a Lagrange multiplier introduced to achieve minimization. The solution of the kriging equations provides the weights in Eq. (4.1), and the prediction variance can be obtained as

$$\sigma^2(\mathbf{x}_0) = \sum_{i=1}^{N} \lambda_i \gamma(\mathbf{x}_i - \mathbf{x}_0) + \psi(\mathbf{x}_0). \tag{4.7}$$

The kriging equations can also be written in matrix form. For punctual kriging they are given by

$$\mathbf{A}\lambda = \mathbf{b}, \tag{4.8}$$

where the matrix $\mathbf{A}$ represents the semivariances between the ith and jth sampling points, λ is the vector of weights and $\mathbf{b}$ is the vector of semivariances between each sampling point and the target point. Matrix $\mathbf{A}$ is inverted, and the weights and Lagrange multipliers are obtained as

$$\lambda = \mathbf{A}^{-1}\mathbf{b}. \tag{4.9}$$

The kriging variance in matrix form is

$$\hat{\sigma}^2(\mathbf{x}_0) = \mathbf{b}^{\mathrm{T}}\lambda. \tag{4.10}$$

The equivalent kriging system for blocks is

$$\sum_{i=1}^{N} \lambda_i \gamma(\mathbf{x}_i - \mathbf{x}_j) + \psi(B) = \bar{\gamma}(\mathbf{x}_j, B) \quad \text{for all } j$$

$$\sum_{i=1}^{N} \lambda_i = 1, \tag{4.11}$$

and the block kriging variance is obtained as

$$\sigma^2(B) = \sum_{i=1}^{N} \lambda_i \bar{\gamma}(\mathbf{x}_i, B) + \psi(B) - \bar{\gamma}(B, B). \tag{4.12}$$

In the matrix representation for block kriging, $\mathbf{b}$ in Eq. (4.8) is the vector of semivariances between each sampling point and the target block and the kriging variance becomes

$$\hat{\sigma}^2(B) = \mathbf{b}^T \lambda - \hat{\gamma}(B, B). \tag{4.13}$$

The variances of block kriging are in general smaller than those of punctual kriging because any nugget variance is contained entirely in the within-block variance, $\bar{\gamma}(B, B)$, and so it disappears from the block-kriging variance, as is evident in Eq. (4.9). There is also somewhat less fluctuation among the block kriged predictions than among punctual ones, and so kriged surfaces made by block kriging are smoother than those made by punctual kriging (see Sect. 4.2.3).

The equations above show clearly the crucial role played by the variogram in kriging, and it is for this reason that we emphasized its importance in Chap. 3.

4.2.1 Kriging Weights

Equations (4.6) and (4.11) show that the kriging weights depend on the variogram; they are functions of the semivariances between the sites in the neighbourhood, $\gamma(\mathbf{x}_i - \mathbf{x}_j)$, and those between each sampling point and the point or block to be predicted, or $\gamma(\mathbf{x}_i - \mathbf{x}_0)$ or $\bar{\gamma}(B, \mathbf{x}_i)$, respectively.

When one solves the kriging equations in practice one usually finds that only points near to the target carry significant weight and that most can be disregarded. Kriging is seen to be a local predictor. Further, it means that only points near to the target need to be included in the kriging systems which can be much smaller than one that includes all N data points and is computationally more tractable. Kriging for mapping then proceeds within a moving window. We illustrate these effects in the following examples. We use a set of data on $\log_{10} K^+$ recorded on a 4×4 grid at Broom's Barn Farm (a farm of 80 ha in Suffolk, England). The best fitting model to these data was a spherical function (see Fig. 3.4a),

$$\gamma(h) = \begin{cases} c_0 + c\left\{ \frac{3h}{2r} - \frac{1}{2}\left(\frac{h}{r}\right)^3 \right\} & \text{for } 0 < h \leq r \\ c_0 + c & \text{for } h > r \\ 0 & \text{for } h = 0, \end{cases} \tag{4.14}$$

in which c_0, the nugget variance, is 0.004, c, the variance of spatially correlated component is 0.016 and r, the range of spatial dependence, is 426 m.

4.2.1.1 Effect of the Ratio of the Nugget:Sill Variances

We have pointed out elsewhere (Oliver and Webster 2014) that the nugget:sill ratio, $c_0: c_0 + c$, obtained from data is strictly a characteristic of the empirical model fitted to an experimental variogram. Because we use the model for kriging, however, the ratio can have important consequences. First for punctual kriging, we show the effect of changing this ratio on the kriging weights, the predictions and the prediction variances. We keep the range constant as for the best fitting spherical function given above. Figure 4.1 shows the layout of the 16 data points with the target at the centre. In Fig. 4.1a the weights are for a variogram model with zero

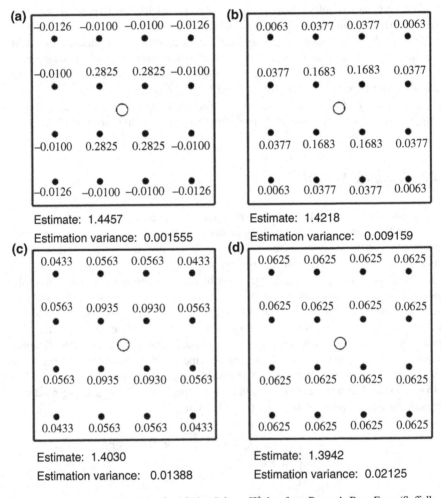

Fig. 4.1 Punctual kriging weights for 16 (4 × 4) $\log_{10}$ K$^+$ data from Broom's Barn Farm (Suffolk, England) based on a spherical function with a range of 426 m. The nugget:sill ratio was changed as follows: **a** $c_0 = 0: c = 0.02$, **b** $c_0 = 0.004: c = 0.016$, **c** $c_0 = 0.012: c = 0.008$ and **d** $c_0 = 02: c = 0$

nugget, a sill of 0.02 and range of 426 m. The four points closest to the target carry almost all of the weight, and the weights of all of the other points are negative. The points at the four corners, which are furthest from the target point, have the smallest absolute weights.

Figure 4.1b shows the weights for the spherical function with a nugget:sill ratio of 0.2. Its distribution of weights differs markedly from that in Fig. 4.1a; the four central points have about 60 % less weight, more weight is given to the next nearest points and to the four corner points, which still carry the smallest weights. Figure 4.1c shows what happens if the nugget:sill ratio is increased to 0.6. The four points closest to the target have much smaller weights than before, whereas the outer points now have considerably larger weights. The overall effect of the larger nugget variance is to weaken the effect of the closest points. Figure 4.1d shows that when the variogram is pure nugget the weights are all the same because all that can be estimated is the mean of these 16 points or of the points in the kriging neighbourhood (see Sect. 4.2.2). Table 4.1 shows how the predictions and prediction variances change with the increasing nugget effect; the variances increase, and in this instance because of the configuration of the data the predicted values decrease. We assume that the most likely value at the point is 1.424 because this was predicted with the best fitting model. Although the estimation variance is smallest for the variogram with zero nugget variance, we should want to be sure that such a model was realistic before placing our trust in the outcome; we should not want the potentially false sense of security that it might give.

These results show how sensitive the weights and prediction variances are to the model parameters and confirm the importance we give in Chap. 3 to computing and modelling the variogram accurately. This is illustrated further in Fig. 4.2 in which the model is anisotropic. The weights near to the target are small in the direction of maximum variation and large in the direction of minimum variation.

If the target point to be predicted is away from the centre of the grid cell the data point closest to it has the largest weight. The weights of the other three central points are still larger than the outer weights, but they are no longer all the same. The latter also applies to the outer weights. Where the target point coincides with a sampling point the weight of the latter is 1 and elsewhere the weights are zero. The predicted value is the value of the datum, and the kriging variance is zero. This illustrates that punctual kriging is an exact predictor, as shown in Fig. 4.10a, b.

Table 4.1 Punctually kriged estimates, estimation variances and Lagrange multipliers determined from 16 $\log_{10}K^+$ data from Broom's Barn Farm, Suffolk, with a spherical model, a set range and different nugget:sill variances

Model properties	Punctual kriging		
Spherical—range 426 m	Kriged estimate	Estimation variance	Lagrange multiplier
$c_0 = 0$; $c = 0.02$	1.446	0.00156	−0.0000436
$c_0 = 0.004$; $c = 0.016$	1.422	0.00599	−0.0002657
$c_0 = 0.008$; $c = 0.012$	1.410	0.00998	−0.0000744
$c_0 = 0.012$; $c = 0.008$	1.403	0.01380	0.0002919
$c_0 = 0.02$; $c = 0$	1.394	0.02125	0.001250

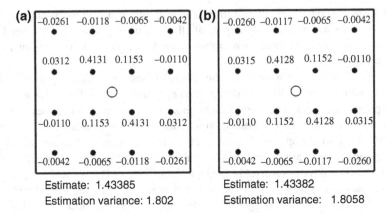

(a)

-0.0261	-0.0118	-0.0065	-0.0042
0.0312	0.4131	0.1153	-0.0110
		○	
-0.0110	0.1153	0.4131	0.0312
-0.0042	-0.0065	-0.0118	-0.0261

Estimate: 1.43385
Estimation variance: 1.802

(b)

-0.0260	-0.0117	-0.0065	-0.0042
0.0315	0.4128	0.1152	-0.0110
		○	
-0.0110	0.1152	0.4128	0.0315
-0.0042	-0.0065	-0.0117	-0.0260

Estimate: 1.43382
Estimation variance: 1.8058

Fig. 4.2 Punctual kriging weights for 16 (4 × 4) $\log_{10}K^+$ data from Broom's Barn Farm (Suffolk, England) based on an anisotropic spherical function with the direction of maximum variation (minimum range), $\varphi = 45°$, a minimum range of 250 m, a maximum range of 480 m and anisotropy ratio of 2. The nugget:sill ratio was changed as follows: **a** $c_0 = 0$:$c = 0.02$ and **b** $c_0 = 0.004$: $c = 0.016$

4.2.1.2 Changing the Range

In this example, we show the effect of changing only the variogram range; the nugget variance, c_0, and spatially dependent component, c, are kept constant with the values of the best fitting spherical model for the data. If the range is much shorter than the sample spacing, for example half (20 m) the sampling interval, then the weights are all the same as for a pure nugget variogram. We see in Table 4.2 that the predicted value and its variance are the same as for the pure nugget example in Table 4.1. Simply, in the model all the variation o ccurs over distances less than the sampling interval. For a range of 120 m, Fig. 4.3a, the weights of the four central points are much larger than the next circle of points and those at the four corners are negative. As the range increases, the weights of the inner circle of points decrease and those of the outer points increase although they are still negative for a

Table 4.2 Punctually kriged estimates, estimation variances and Lagrange multipliers determined from 16 $\log_{10}K^+$ data from Broom's Barn Farm, Suffolk, with a spherical model, a set nugget:sill variance and different ranges

Model properties	Punctual kriging		
Spherical—$c_0 = 0.004$; $c = 0.016$ (m)	Kriged estimate	Estimation variance	Lagrange multiplier
Range, $r, = $ 20	1.394	0.02125	0.001250
Range, $r, = 120$	1.439	0.00959	0.0000634
Range, $r, = 280$	1.428	0.00676	-0.000308
Range, $r, = 426$	1.422	0.00599	-0.000266
Range, $r, = 680$	1.416	0.00541	-0.001682

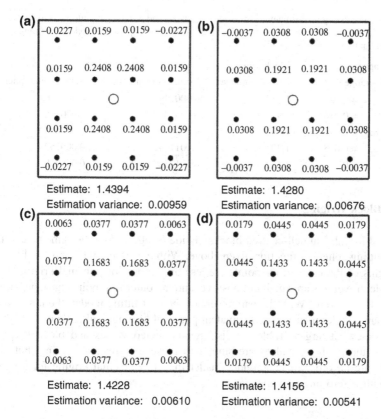

(a)

−0.0227	0.0159	0.0159	−0.0227
0.0159	0.2408	0.2408	0.0159
0.0159	0.2408	0.2408	0.0159
−0.0227	0.0159	0.0159	−0.0227

Estimate: 1.4394
Estimation variance: 0.00959

(b)

−0.0037	0.0308	0.0308	−0.0037
0.0308	0.1921	0.1921	0.0308
0.0308	0.1921	0.1921	0.0308
−0.0037	0.0308	0.0308	−0.0037

Estimate: 1.4280
Estimation variance: 0.00676

(c)

0.0063	0.0377	0.0377	0.0063
0.0377	0.1683	0.1683	0.0377
0.0377	0.1683	0.1683	0.0377
0.0063	0.0377	0.0377	0.0063

Estimate: 1.4228
Estimation variance: 0.00610

(d)

0.0179	0.0445	0.0445	0.0179
0.0445	0.1433	0.1433	0.0445
0.0445	0.1433	0.1433	0.0445
0.0179	0.0445	0.0445	0.0179

Estimate: 1.4156
Estimation variance: 0.00541

Fig. 4.3 Punctual kriging weights for 16 (4 × 4) $\log_{10}K^+$ data from Broom's Barn Farm (Suffolk, England) based on a spherical function with a nugget variance, $c_0 = 0.004$ and sill variance, $c = 0.016$. The range was changed as follows: **a** $a = 120$ m, **b** $a = 280$, **c** $a = 426$ m and **d** $a = 680$

range of 280 m, Fig. 4.3b. For the two longest ranges in this example, 426 m, Fig. 4.3c and 680 m, Fig. 4.3d, the inner weights become smaller and the outer ones larger. The predicted values have no clear pattern, and, apart from the very short range, the values are similar to one another. The prediction variances decrease with the increase in range, which reflects the implied increase in continuity of the process. These results show that if the nugget variance is estimated correctly then the range is less important for kriging. This is because kriging uses only semivariances near to the ordinate of the variogram. However, for sample design and interpretation of the variation, it is important to ensure that the range is also estimated precisely (see Chap. 5).

Table 4.3 Block kriged estimates, estimation variances and Lagrange multipliers determined from 16 $\log_{10}K^+$ data from Broom's Barn Farm, Suffolk, with a spherical model, a set range and different nugget:sill variances

Model properties	Block kriging		
Spherical—range 426 m	Kriged estimate	Estimation variance	Lagrange multiplier
$c_0 = 0$; $c = 0.02$	1.437	0.000149	−0.000050
$c_0 = 0.004$; $c = 0.016$	1.417	0.000706	−0.000196
$c_0 = 0.008$; $c = 0.012$	1.407	0.000960	−0.0000079
$c_0 = 0.012$; $c = 0.008$	1.401	0.001104	0.0003590
$c_0 = 0.02$; $c = 0$	1.394	0.001250	0.001250

4.2.1.3 Block Kriging

We kriged $\log_{10}K^+$ in cell-centred blocks of side of 60 m from the same data with the best fitting spherical function given above. With $c_0 = 0$, $c = 0.02$ and $r = 426$ m, the weights of the centre four points are less than those for punctual kriging, the next circle of weights are small but positive, and the outer four points have negative weights close to zero. With the parameters of the best fitting model, the inner four weights are smaller than for the equivalent punctual kriging, and the outer weights are all somewhat larger. Table 4.3 lists the predicted values and block kriging variances; the predicted values are similar to those for punctual kriging, but the variances are an order of magnitude smaller than for punctual kriging for a variogram with a zero nugget variance.

4.2.1.4 Kriging with Irregularly Scattered Data

In the final example, we selected 11 irregularly distributed sites and used the best fitting model for punctual kriging. Figure 4.4 shows the distribution of the 11 data points and the target point to be estimated. The weights for the variogram with zero nugget have a markedly different distribution from those on the grid, Fig. 4.4a. The closest point to the target has a substantially larger weight than the other closest points. The distribution of the weights, however, illustrates some other effects on the weights. The weight of the point to the north east of the target is smaller than that of the point to its left—this is because it is somewhat further from the target and its weight is also affected by the proximity of this point. The point to the north of it has a negative weight because it is in the shadow of the nearer point. The point to the east of the target has a larger weight than that to the north east of the target although it is further from the target. This is because no other points lie close to it. The outermost point in the south west corner has a larger weight than that to its north because it is not in the shadow of any other points, whereas the point in the north east corner has a negative weight because it is in the shadow of another. Figure 4.4b shows the result for the best fitting spherical function to the $\log_{10}K^+$ data. There is still a marked difference between the closest point to the target and

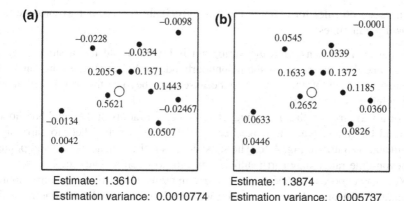

Fig. 4.4 Punctual kriging weights for 11 irregularly scattered $\log_{10}K^+$ data from Broom's Barn Farm (Suffolk, England) based on a spherical function with a range, a, of 426 m and nugget:sill variance: **a** $c_0 = 0$:$c = 0.02$ and **b** $c_0 = 004$:$c = 0.016$

the next closest, but these weights are considerably smaller. Of interest is the weight of the point to the east of the target—this is now smaller than that of the point to the north east. The effect of the nugget variance is to even out the weights, and this effect appears greater for irregularly spaced data than for the grid. The results for the irregularly spaced data emphasize the greater need for accuracy in computing a variogram from irregularly spaced data (see also Chap. 3) (Fig. 4.4).

4.2.2 Kriging Neighbourhood

We pointed out that kriging is essentially local, and the above examples show that it is so provided the nugget:sill ratio of the model is small. The matter does not rest there, however.

The local nature of ordinary kriging means that we can accept the assumption of local stationarity (or quasi stationarity), which means that we can restrict the assumption of stationarity of the mean to that within the kriging neighbourhoods. What happens over distances larger than that across the neighbourhood is of little consequence for the predictions, and the underlying assumptions of the method are not violated in practice. Another aspect of the matter is of greater significance; the kriged predictions and especially their associated variances depend very much on that part of the variogram model close to the ordinate, i.e. over lag distances shorter than the distances between target points and their nearest neighbours. This means that you should estimate and model the variogram well there. It supports the idea of giving more weight to the experimental semivariances near to the origin when you model the variogram specifically for kriging (see Chap. 3).

There are no rules for defining the kriging neighbourhood, but we provide the following guidelines.

1. If the data are dense and the variogram is bounded and has a small nugget variance then the radius of the neighbourhood can be set close to the range or effective range of the variogram because data beyond the range will have negligible weight.
2. For a variogram with a large nugget variance, the radius of the neighbourhood could be greater than the range because distant points are likely to carry significant weight, see Fig. 4.1c. The same applies if the data are sparse and points beyond the range may carry sufficient weight to be important (see Fig. 4.4).
3. You may choose to set the neighbourhood in terms of a minimum and maximum number of nearest data to the target; we usually recommend a minimum $n \approx 7$ and a maximum $n \approx 20$. The minimum is needed for targets near to the boundary of the region.
4. If the data are very unevenly scattered it is good practice to divide the neighbourhood into octants so that there are at least two points in each. The good sense of this is evident in Fig. 4.4.
5. When you start to analyse new data examine what happens to the kriging weights as you change the neighbourhood, especially where the data are irregularly scattered and you use a moving window for mapping.

4.2.2.1 Effect of the Kriging Neighbourhood

To illustrate the effect of the kriging neighbourhood, we use an example of readily extractable cobalt (Co) in the topsoil from the Borders Region of Scotland. The data are from an original study by McBratney et al. (1982). There were almost 2 000 measured concentrations in the eastern part of the region, and the values in mg kg^{-1} were transformed to their common logarithms to reduce the skewness and so stabilize the variances. Table 4.4 summarizes the statistics of the $\log_{10}$Co data.

Table 4.4 Summary statistics of cobalt and $\log_{10}$Co from the Borders Region of Scotland		Co/mg kg^{-1}	$\log_{10}$Co
	Number of data	1980	1980
	Minimum	0.050	−1.301
	Maximum	1.00	0
	Mean	0.254	−0.639
	Median	0.220	−0.658
	Standard deviation	0.123	0.195
	Variance	0.0151	0.0379
	Skewness	1.606	0.140

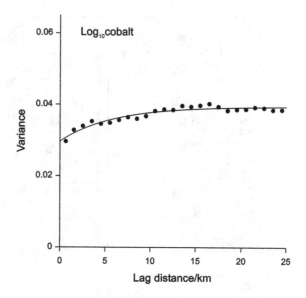

Fig. 4.5 Experimental variogram of $\log_{10}Co$ (*symbols*) in the Borders Region of Scotland with a fitted exponential model (*solid line*)

An isotropic experimental variogram was computed by the method of moments, and it was fitted by an isotropic exponential function which was used for kriging. The exponential function is given by

$$\gamma(h) = c_0 + c\left\{1 - \exp\left(-\frac{h}{a}\right)\right\}, \qquad (4.15)$$

where c_0 is the nugget variance, c is the variance of spatially correlated component and r is the distance parameter. Table 4.6 lists the model parameters and Fig. 4.5 shows the variogram; the symbols are the experimental values and the solid line is the fitted model. The prediction grid was 100 m × 100 m, and the size of the moving kriging neighbourhood (or window) was varied as follows to include: (1) a maximum of 20 points, a minimum of 17 points; (2) maximum of 40, minimum of 37; (3) maximum of 80, minimum of 77 and (4) maximum of 140, minimum of 137. It is usual to set a window size with a much larger maximum number of points than the minimum, but here we have used a larger minimum to ensure that we know how many points are in the neighbourhood. Figure 4.6 shows the result of mapping the kriged predictions of $\log_{10}Co$ for the various neighbourhood sizes. Figure 4.6a for the smallest neighbourhood size shows detail that is not apparent in the maps for the larger neighbourhoods. Figure 4.6b shows more variation than Fig. 4.6c, d; these latter maps show a similar degree of variation. As the neighbourhood size increases, the detail decreases to a point at which it becomes stable.

Figure 4.7 shows the maps of kriging variances for the four neighbourhood sizes. The kriging variances are largest for the smallest neighbourhood of 20 points, Fig. 4.7a, and they decrease considerably for the second neighbourhood size of 40 points, Fig. 4.7b, and by a decreasing amount for the third neighbourhood size of 80

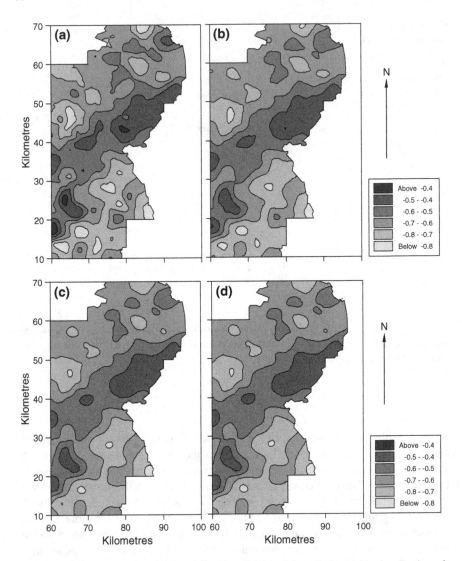

Fig. 4.6 Maps of ordinary punctually kriged predictions of $\log_{10}Co$ in the Borders Region of Scotland with the kriging neighbourhood set as follows: **a** 20 maximum points, 17 minimum points, **b** 40 maximum points, 37 minimum points, **c** 80 maximum points, 77 minimum points and **d** 140 maximum points, 137 minimum points

points, Fig. 4.7c. There is no further decrease in the kriging variances for the largest kriging neighbourhood of 140 points, Fig. 4.7d.

Evidently automation without regard to the nugget effect can produce unwarranted detail. The larger is the nugget:sill ratio in your model the larger should be the neighbourhood within which to search for data.

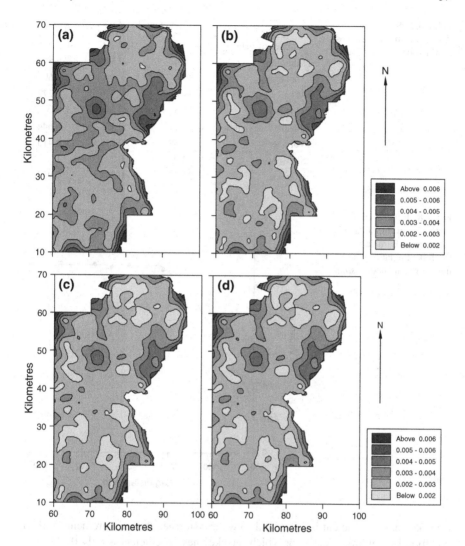

Fig. 4.7 Maps of ordinary punctual kriging variances of $\log_{10}$Co in the Borders Region of Scotland with the kriging neighbourhood set as follows: **a** 20 maximum points, 17 minimum points, **b** 40 maximum points, 37 minimum points, **c** 80 maximum points, 77 minimum points and **d** 140 maximum points, 137 minimum points

4.2.3 Punctual and Block Kriging for Mapping

We have already distinguished punctual and block kriging. We illustrate the difference with data from a case study by Webster and McBratney (1987) on available phosphorus, P, in the topsoil of Broom's Barn Farm. Table 4.5 summarizes the data from the survey at 40-m intervals on a square grid. The data are strongly skewed

Table 4.5 Summary statistics of three variables from Broom's Barn Farm

	P/mg l^{-1}	Log$_{10}$(P)	pH
Number of data	433	434	435
Minimum	0.4	−0.398	5.50
Maximum	49.0	1.69	8.60
Mean	4.865	0.546	7.272
Median	3.5	0.544	8.0
Standard deviation	5.150	0.338	0.644
Variance	26.52	0.114	0.415
Skewness	3.946	0.231	−1.30
Kurtosis	21.47	0.364	0.952

Fig. 4.8 Experimental variogram of log$_{10}$P (*symbols*) at Broom's Barn farm (Suffolk, England) with a fitted circular model (*solid line*)

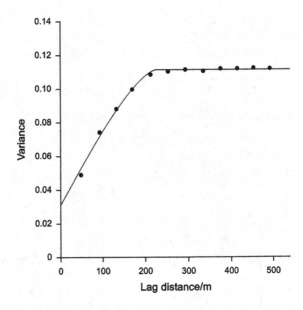

(the skewness coefficient is 3.95), and so we transformed the measurements to their common logarithms, log$_{10}$P, for which the skewness coefficient is only 0.364. The variogram was computed on the transformed data, and the experimental semi-variances were fitted best by a circular model given by:

$$\gamma(h) = \begin{cases} c_0 + c\left\{1 - \frac{2}{\pi}\cos^{-1}\left(\frac{h}{r}\right) + \frac{2h}{\pi r}\sqrt{1 - \frac{h^2}{r^2}}\right\} & \text{for } 0 < h \le r, \\ c_0 + c & \text{for } h > r, \\ 0 & \text{for } h = 0. \end{cases} \quad (4.16)$$

Figure 4.8 shows the experimental variogram (symbols) and the fitted circular model (solid line), and Table 4.6 lists the model parameters. The circular function was then used for both punctual and block kriging. We set the maximum radius of the neighbourhood to 380 m and the minimum number of points in the

Table 4.6 Parameters of models fitted to readily extractable cobalt in the Borders Region of Scotland and exchangeable phosphorus and pH in the topsoil at Broom's Barn Farm, Suffolk, England

Model type	Estimates of parameters							
	c_0	c	r/m	a/m	φ (rad)	A/m	B/m	A/B
Exponential—$\log_{10}$Co	0.02967	0.009784		5750.0				
Circular—$\log_{10}$P	0.03103	0.08036	225.8					
Exponential—pH	0	0.3588		85.2				
Anisotropic exponential—pH	0	0.3710			2.735 (156.7°)	126.9	61.96	2.048

neighbourhood to seven and the maximum to 20. We kriged at the nodes of a 10-m grid. For block kriging we chose blocks of 20 m × 20 m.

Provided the measurements are free of error, punctual kriging at sampling points returns the data values there with a variance of zero. Figure 4.9a shows the kriged map that includes predictions at sampling points. It appears spotty because of the isarithms, 'contours', that surround many of the sampling points. The effect arises largely because of the nugget variance, which is 28 % of the sill variance. The nugget variance represents a discontinuity, which is present in the kriged results. Figure 4.10a is a perspective diagram of these punctually kriged estimates; the spikes above and below the surface at many of the sampling sites illustrate the discontinuity. The spikes represent the measured values at the sampling points; elsewhere the predictions are weighted averages of data in the neighbourhoods. The kriged estimates effectively comprise two parts, namely an uncorrelated component represented by the nugget variance, and the spatially correlated component, represented by c in Eq. (4.12). The larger is the nugget variance in relation to the total variance, the greater is this effect. If all the variance were nugget, the surface would become flat between sampling points because punctual kriging simply returns the means of the data within the neighbourhood.

Figure 4.9c is a map of the punctual kriging variances. Because the prediction grid coincides with the data grid, the values at the sampling points are zero. In the perspective diagram, Fig. 4.10b, they are represented by holes in the surface descending to zero. The perspective diagram shows another effect of punctual kriging, i.e. that the surface appears to be sitting on a platform. This is the nugget variance; the minimum possible kriging variance with punctual kriging away from the data points is set by the nugget variance.

We also examine the effects of punctual kriging on a prediction grid that was offset from the sampling grid. For practical situations we should avoid kriging at the data points because the values are already known there and because of the adverse effects that arise when the variogram has a nugget variance as mentioned above. The resulting map from prediction with the grid offset from the sampling points is smoother, Fig. 4.9b; it no longer appears spotty. The perspective diagram, Fig. 4.10c, no longer has spikes at the data points; it is more like that for block

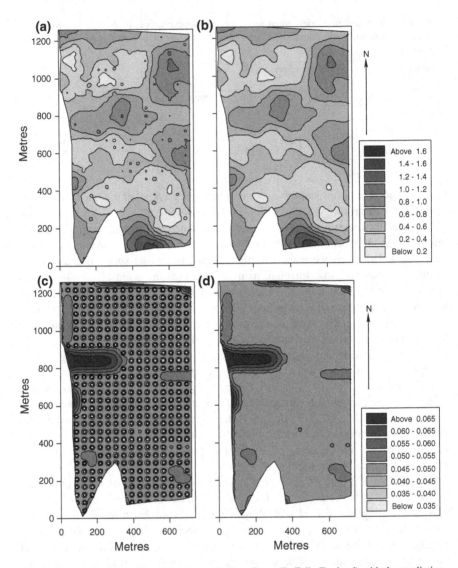

Fig. 4.9 Punctual kriging of $\log_{10}P$ at Broom's Barn Farm (Suffolk, England) with the prediction grid coinciding with the sampling grid: **a** kriged predictions and **c** kriging variances, and with the prediction grid offset from the sampling grid: **b** kriged predictions and **d** kriging variances

kriging, Fig. 4.10e. The map of the kriging variances, which has been plotted with the same scale of values as for the grid that coincided with the sampling grid, Fig. 4.9d, has lost the mattress-like appearance of Fig. 4.9c and the errors appear larger. We advise you to offset your prediction grid to get a clear impression of the variation without the distractions of discontinuities in the surface. Figure 4.10d, the perspective diagram of the kriging variances with the grid offset is similar to that in

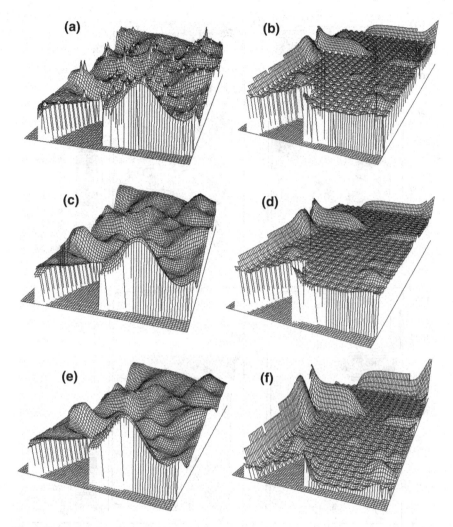

Fig. 4.10 Perspective diagrams of $\log_{10}P$ at Broom's Barn Farm (Suffolk, England): **a** punctually kriged predictions, prediction and sampling grids coincide, **b** punctual kriging variances, prediction and sampling grids coincide, **c** punctually kriged predictions, prediction and sampling grids are offset, **d** punctual kriging variances, prediction and sampling grids are offset, **e** block kriged predictions, prediction and sampling grids coincide and **f** block kriging variances, prediction and sampling grids coincide

Fig. 4.10b—the effect of the nugget variance remains in the kriging variances despite the grid's being offset.

We also kriged over blocks of side 20 m with the prediction grid positioned to coincide with the sampling grid. The kriged estimates are similar to those from

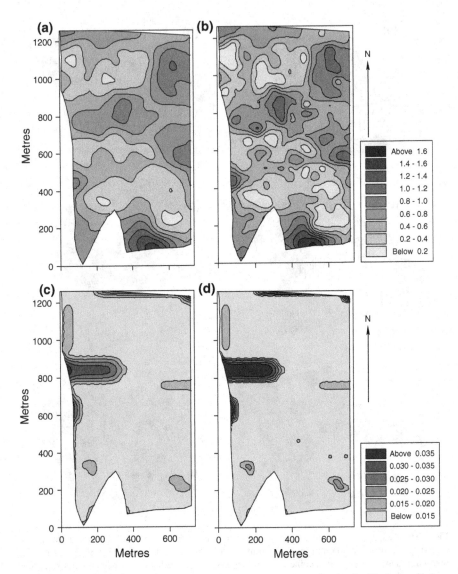

Fig. 4.11 Block kriging of $\log_{10} P$ at Broom's Barn Farm (Suffolk, England) with the prediction grid coinciding with the sampling grid and the correct model function: **a** kriged predictions and **c** kriging variances, and with the variogram forced through the origin to give zero nugget variance **b** kriged predictions and **d** kriging variances

punctual kriging, but are smoother, as shown by the map in Fig. 4.11a and perspective diagram, Fig. 4.10e. Such a map is one a farmer would want for managing fertilizer applications.

Farmers are not the only people who want block kriged predictions. For miners block kriging is crucial because they must estimate the grades of blocks of ore from small core samples. Agencies responsible for managing rivers and water supply need to estimate rainfall over portions of their catchments from rain gauges. Block kriged predictions are required also by those agencies responsible for cleaning up contaminated sites.

Although the predictions are similar to those of punctual kriging, the kriging variances are now substantially smaller, Fig. 4.11c, d. This implies that the predictions are more precise than are those from punctual kriging. In block kriging, the nugget variance disappears from the block-kriging variance because it is contained in the within-block variance. This is clear from the perspective diagram of the block kriging variances; the platform representing the nugget variance in Fig. 4.10b, d has now disappeared as the theory indicates.

Before we leave this matter we should clarify what the nugget variance is. In theory it is a spatially uncorrelated component of variation; it is white noise in engineering terms. In environmental science the variables of interest, such as the pH and phosphorus content of the soil, are continuous; there is no uncorrelated component. In practice, however, we can rarely observe the environment continuously; we have to sample to determine quantities such as the pH and phosphorus content of the soil. We cannot therefore know the form of the variogram as the lag distance approaches zero from the shortest sampling interval. Instead we take a conservative approach: we fit models to the ordered set of calculated semivariances and extrapolate them to the ordinate to obtain the nugget variance. This nugget variance therefore consists largely of variance within the shortest sampling interval; it is largely not the variance of white noise. It does include measurement error, which may be regarded as white noise, but its variance should be much smaller than the spatial variance.

A consequence of this practice is that the errors of punctual kriging are likely to be somewhat exaggerated, whereas those of block kriging are likely to be underestimated for the reasons mentioned above. If you wish to improve matters then you need to elaborate the sampling scheme with numerous closely spaced observations in addition to any coarse grid, as described by Webster and Lark (2013) and in Chap. 5. Only by so doing can you estimate the form of the variogram near the ordinate.

4.2.4 Anisotropy

We now examine the effect of anisotropy with an example, again from Broom's Barn Farm where the pH of the soil was measured. The variogram of pH appears anisotropic, as shown in Fig. 3.12. We fitted the anisotropic exponential model (the model parameters are given in Table 4.6), and we block kriged with this function (Fig. 4.12).

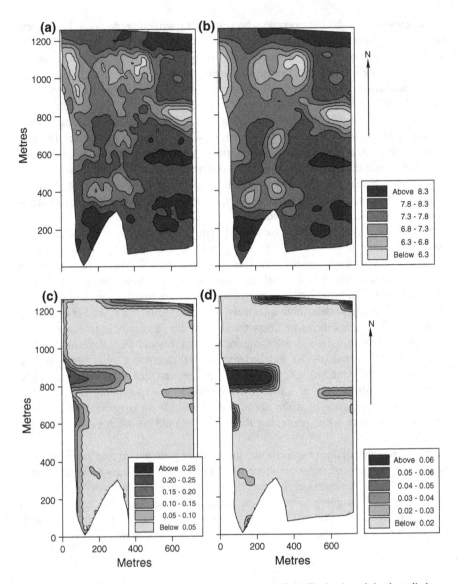

Fig. 4.12 Block kriging of pH at Broom's Barn Farm (Suffolk, England): **a** kriged predictions with the correct anisotropic exponential function, **b** kriged predictions with the poorer fitting isotropic exponential function, **c** kriging variances with the anisotropic variogram and **d** kriging variances with the isotropic function

$$\gamma(h, \vartheta) = c_0 + c \left\{ 1 - \exp\left(\frac{|h|}{\Omega(\vartheta)} \right) \right\}, \qquad (4.17)$$

where $\Omega(\vartheta)$ is

$$\Omega(\vartheta) = \left\{ A^2 \cos^2(\vartheta - \varphi) + B^2 \sin^2(\vartheta - \varphi) \right\}^{1/2}. \tag{4.18}$$

Figure 4.12a shows the result; the anisotropy is evident with more variation in a south west to north east direction than in the perpendicular direction, especially in the northern part of the farm. We also kriged with the less well-fitting isotropic exponential model; the model parameters are given in Table 4.6. The mapped estimates made with this function have a similar general pattern, but with some loss in detail, Fig. 4.12b. The maps of the kriging variances appear similar, but those for the anisotropic function, Fig. 4.12c, are a magnitude larger than are those from the isotropic function, Fig. 4.12d. These results again emphasize the importance of choosing the most appropriate model for prediction; the results from the isotropic function give a false sense of security of the accuracy of the predictions.

4.2.5 Simple Kriging

Ordinary kriging has become the 'workhorse' of geostatistical prediction with its assumption of intrinsic stationarity easily satisfied and its robustness in the face of departures from that assumption. There are circumstances in which we can draw on the somewhat stronger assumption of second-order stationarity of constant mean, which we might know or can assume. In these circumstances we can improve our predictions by *simple kriging*. The kriged predictions are still linear sums of the data, but now they incorporate the mean, μ. The equation for simple punctual kriging is

$$\hat{Z}(\mathbf{x}_0) = \sum_{i=1}^{N} \lambda_i z(\mathbf{x}_i) + \left\{ 1 - \sum_{i=1}^{N} \lambda_i \right\} \mu. \tag{4.19}$$

The weights, λ_i, are the weights as in ordinary kriging, but they are no longer constrained to sum to 1. Inclusion of the second term on the right-hand side of Eq. (4.19) ensures that the predictions are unbiased. Another change is that we have to work with covariances, C, rather than semivariances in these circumstances. The equations to be solved as expressed in the simple punctual kriging system are

$$\sum_{i=1}^{N} \lambda_i C(\mathbf{x}_i, \mathbf{x}_j) = C(\mathbf{x}_0, \mathbf{x}_j) \qquad \text{for } j = 1, 2, \ldots, N \tag{4.20}$$

There is no Lagrange multiplier because the mean is assumed to be known; there are only N equations in N unknowns.

The simple kriging variance is given by

$$\sigma_{SK}^2(\mathbf{x}_0) = C(\mathbf{0}) - \sum_{i=1}^{N} \lambda_i C(\mathbf{x}_i, \mathbf{x}_0), \qquad (4.21)$$

where $C(\mathbf{0})$ is the variance of the random process.

The simple kriging system can be elaborated for predicting values over blocks, B, larger than the supports of the data; all we have to do is to replace the covariance $C(\mathbf{x}_0, \mathbf{x}_j)$ on the right-hand side of Eq. (4.20) by the average covariance $C(B, \mathbf{x}_j)$.

As in ordinary kriging, we can usually replace N by some much smaller value n by kriging within a moving window for mapping. Simple kriging, whether punctual or block, is somewhat more precise than its ordinary counterpart because there is no longer any uncertainty in the mean, and it is in that sense that it is an improvement. However, simple kriging is limited by the need to know the mean.

4.2.6 Lognormal Kriging

Environmental data often approximate a lognormal distribution. Converting the data to logarithms produces distributions that are approximately normal and this leads to lognormal kriging. The data $z(\mathbf{x}_1)$, $z(\mathbf{x}_2)$, ..., are transformed to their corresponding natural logarithms, say $y(\mathbf{x}_1)$, $y(\mathbf{x}_2)$, ..., which represent a sample from the random variable $Y(\mathbf{x}) = \ln Z(\mathbf{x})$, which is assumed to be second-order stationary. The variogram of $Y(\mathbf{x})$ is computed and modelled and then used with the transformed data to predict Y at the target points or blocks by either ordinary or simple kriging.

The predictions are logarithms, and where an index of soil fertility is wanted the logarithms can serve well. However, in many other disciplines, such as mining, exploration geochemistry and pollution monitoring, surveyors want estimates expressed in the original units and the logarithms must be transformed back to concentrations.

The back-transformation of a punctual estimate is fairly straightforward. If we denote the kriged estimate of the natural logarithm at $\mathbf{x}_0$ as $\hat{Y}(\mathbf{x}_0)$ and its variance as $\sigma^2(\mathbf{x}_0)$ then the formulae for the back-transformation of the estimates are, for simple kriging,

$$\hat{Z}_{SK}(\mathbf{x}_0) = \exp\{\hat{Y}_{SK}(\mathbf{x}_0) + \sigma_{SK}^2(\mathbf{x}_0)/2\}, \qquad (4.22)$$

and for ordinary kriging,

$$\hat{Z}_{OK}(\mathbf{x}_0) = \exp\{\hat{Y}_{OK}(\mathbf{x}_0) + \sigma_{SK}^2(\mathbf{x}_0)/2 - \psi(\mathbf{x}_0)\}, \qquad (4.23)$$

where ψ is the Lagrange multiplier in ordinary kriging. The estimation variance of $\hat{Z}(\mathbf{x}_0)$ for simple kriging is

$$\text{var}_{\text{SK}}[\hat{Z}(\mathbf{x}_0)] = \hat{Z}^2(\mathbf{x}_0)\exp\{\sigma^2_{\text{SK}}(\mathbf{x}_0)\}\,[1 - \exp\{-\sigma^2_{\text{SK}}(\mathbf{x}_0)\}]. \tag{4.24}$$

Notice that the distribution around each prediction now depends on the prediction itself and therefore on the observed values in the neighbourhood. Other things' being equal, the larger are those observed values the larger will be the prediction variance. See also Papritz and Moyeed (1999) and, for further details, Cressie (1993), pp. 135 and 136.

In many fields of application people prefer to work with common logarithms. The variogram of $\log_{10} Z(\mathbf{x})$ replaces that of $\ln Z(\mathbf{x})$, and the back-transform for ordinary kriging, is given by

$$\hat{Z}(\mathbf{x}_0) = \exp[\{\hat{Y}(\mathbf{x}_0) \times \ln 10 + 0.5\sigma^2_Y(\mathbf{x}_0) \times (\ln 10) - \psi(\mathbf{x}_0) \times (\ln 10)\}]. \tag{4.25}$$

Journel and Huijbregts (1978) point out that the expression in Eqs. (4.22) and (4.23) for the back-transformation, is sensitive to departures from lognormality and that in consequence the estimates of Z can be biased. They suggest a check for bias by comparing the mean of the estimates, $\hat{Z}$, with the mean of the data, $z(\mathbf{x}_i)$, $i = 1, 2, \ldots, N$. If we denote the ratio of the means, $\text{mean}[\hat{Z}]{:}\bar{z}$, by Q then we modify Eq. (4.22) to

$$\hat{Z}_{\text{SK}}(\mathbf{x}_0) = Q \exp\{\hat{Y}_{\text{SK}}(\mathbf{x}_0) + \sigma^2_{\text{SK}}(\mathbf{x}_0)/2\}, \tag{4.26}$$

or Eq. (4.23) similarly if we have used ordinary kriging. In our experience Q has always been so close to 1 that we have not needed the elaboration. For more information see Cressie (2006).

4.3 Cross-Validation

The examples above show how the choice of model for a variogram affects the kriging weights and the kriging variances, even though the kriged predictions are little affected. Further, given that several models with substantially different nugget: sill ratios can appear to fit an experimental variogram equally well, one might wonder how to choose the most suitable for kriging. Minimum variance is attractive, but we do not want that if it leads to a false sense of security. Rather, we want a model that leads to variances that are 'correct' in the sense that they match the squared errors between the predictions and the true values. One could compare competing models by predicting values with those observed at an independent set of sampling points. In practice that would waste valuable, possibly very expensive, information. The alternative is to make the comparisons by cross-validation.

Let us assume that we have two or more plausible models. For each model we proceed as follows. We remove one datum from the whole set and use either all the others or those in the surrounding neighbourhood and the parameters of the given

model to krige its value and calculate its kriging variance. We return that datum to
the set, and repeat the procedure for each and every one of the remaining data. The
kriged values $\hat{Z}(\mathbf{x}_i)$ are compared with the observed ones $z(\mathbf{x}_i)$, and we then calculate the following statistics from the results.

1. The mean deviation or mean error, ME:

$$\text{ME} = \frac{1}{N} \sum_{i=1}^{N} \{z(\mathbf{x}_i) - \hat{Z}(\mathbf{x}_i)\}, \tag{4.27}$$

where N is the number of observations, $z(\mathbf{x}_i)$ is the true value at $\mathbf{x}_i$ and $\hat{Z}(\mathbf{x}_i)$ is
the predicted value there;
2. The mean squared deviation or mean squared error, MSE:

$$\text{MSE} = \frac{1}{N} \sum_{i=1}^{N} \{z(\mathbf{x}_i) - \hat{Z}(\mathbf{x}_i)\}^2; \tag{4.28}$$

3. The mean squared deviation ratio, MSDR, computed from the squared errors
and the kriging variances, $\hat{\sigma}^2(\mathbf{x})$;

$$\text{MSDR} = \frac{1}{N} \sum_{i=1}^{N} \frac{\{z(\mathbf{x}_i) - \hat{Z}(\mathbf{x}_i)\}^2}{\hat{\sigma}^2(\mathbf{x}_i)}. \tag{4.29}$$

The ME should ideally be 0: kriging is unbiased. The calculated value is a weak
diagnostic because kriged values are insensitive to changes in the model, as we
have seen. We might want to minimize the MSE in the same way as we krige to
minimize the prediction variance. We should like the MSE to be small, but it is not
an especially good diagnostic; it does not discriminate as well as the MSDR.
Ideally, we should like the squared errors to equal their corresponding kriging
variances, so that the MSDR is 1. The best model in this sense is the one for which
the MSDR is most nearly 1.

The results of cross-validation do not necessarily resolve or justify a choice of
model. In the example in Chap. 3 in 3.4.1, the MSDR is closest to 1 for the
spherical models with ranges well beyond the limit of the experimental variogram,
yet Fig. 3.6d strongly suggests that the best fitting model is spherical with
$r \approx 10$ km

Table 4.7 records the calculated values of the criteria for the spherical, exponential and power models fitted to the experimental variograms of models of
$\log_{10}K^+$ at Broom's Barn Farm listed in Table 3.4. The mean errors are all small in
accord with expectation. The mean squared errors are remarkably similar for both
the full set and subset of the data and do not discriminate between the models. The
MSDRs, in contrast, do discriminate. The spherical model fitted to the full set of
data is clearly good for kriging in that its MSDR is close to 1. The exponential and

Table 4.7 Cross-validation criteria, the mean error (ME), the mean squared error (MSE) and the mean squared deviation ratio (MSDR) for models listed in Table 3.5 fitted to the experimental variograms of $\log_{10}K^+$ at Broom's Barn

Model type	Full set, 434 data			Subset, 87 data		
	ME	MSE	MSDR	ME	MSE	MSDR
Spherical	0.000362	0.007614	1.031	0.000659	0.007961	0.732
Exponential	0.00776	0.007349	1.197	0.000957	0.007316	0.876
Power	0.000682	0.007556	1.559	0.000836	0.007825	0.646

power models for the full set have MSDRs that exceed 1, meaning that their kriging variances underestimate the squared errors, and are less good. The MSDRs for all three models fitted to the experimental variograms of the subset of 87 data are substantially less than 1. Evidently the kriging variances overestimate the squared errors, and the models are poor for kriging therefore. This should serve as a further warning against sampling too sparingly.

Even if one chooses a model for which the MSDR is close to 1 for kriging the kriging variances can provide optimistic assessments of the reliability of predictions, especially in the neighbourhood of outliers, and compensate by under-estimating the reliability elsewhere. In such a situation the distribution of the deviations, $z(\mathbf{x}_i) - \hat{Z}(\mathbf{x}_i)$, has long tails; the distribution is leptokurtic, and this property can be identified by the median of the squared deviation ratio: medSDR. Lark (2000) has suggested that the medSDR be used as a diagnostic. Ideally the SDR is distributed as χ^2 with one degree of freedom and its median has the value 0.455. Smaller values suggest leptokurtosis and the presence of outliers—see Li et al. (2015) for an example. Medians exceeding 0.455 mean that the kriging variances underestimate the squared errors, medians less than 0.455 mean that they overestimate. Our experience with this diagnostic has been mixed, mainly because the actual distributions have been erratic. If you intend to use it then we recommend that you examine your distributions of the errors.

4.4 Summary

Early workers in the field were held back by the lack of computer power, but this is no longer the case. Investigators must understand the need for a reliable and well-modelled variogram. Ensure that you know at the outset whether you require punctual or block kriging; the latter is usually chosen for land management, whereas the former would be desirable where the values at points are required. The kriging neighbourhood must be set in relation to what you know about the variation.

Chapter 5
Sampling

Abstract Reliable analysis and kriging demand sound sampling, which must be sufficient and have an acceptable configuration. Sampling to estimate the variogram is problematic because the spatial scale of variation is often unknown, yet there must be numerous pairs of sampling points within the correlation range, if it exists. One might determine the spatial scale from visible features such as landforms and vegetation on the ground or from remote sensing. If that is not possible a nested survey and hierarchical analysis, by either analysis of variance or residual maximum likelihood (REML), can provide a first approximation to the variogram and a guide for subsequent sampling. Variograms from previous surveys or from ancillary data, in particular aerial image data, may also be used to guide sampling. Once a variogram with known parameters is available sampling for kriging can be optimized so that some tolerable kriging error is met but never exceeded. Alternatively, if the budget for sampling is set the kriging equations can be solved to determine the kriging errors everywhere within the region of interest and in particular the maximum absolute error.

Keywords Model-based sampling · Spatial scale · Ancillary information · Nested sampling · Hierarchical analysis of variance · REML · Kriging variance · Optimal sampling · Tolerable error · Mapping

In Chap. 1 we introduced the need for sampling of the environment because of the extent of area usually covered and because the variation is usually continuous. We mentioned design-based and model-based approaches, and here we focus on the latter where our principal concern will be to sample adequately and without bias to enable us to predict accurately throughout the region. This requires sample data that are suitable to estimate both the variogram and to krige. If one knows the variogram of a variable for a particular region and can specify the maximum tolerable error in predictions using it then one can optimize one's sampling scheme (see Sect. 5.2). In most instances, however, one must first estimate the variogram, and we therefore describe the associated problems and the way to tackle them before dealing with the kriging.

M.A. Oliver and R. Webster, *Basic Steps in Geostatistics: The Variogram and Kriging*, SpringerBriefs in Agriculture, DOI 10.1007/978-3-319-15865-5_5

5.1 Sampling for the Variogram

Sampling to estimate the variogram is one of the most problematic tasks in geo-statistics. It receives too little attention among both research workers and practi-tioners with the result that in many instances the data are too few or the spacings are unsuitable for reliable estimates of the variogram. There have been several attempts to optimize sampling for variograms, but without knowing the true variogram one cannot succeed. Lark (2002) and Webster and Lark (2013) show that without prior information on a variogram's likely form and model parameters designing a sam-pling scheme is little better than guesswork. In particular, one must guess the limit of spatial dependence, if such exists in the region. In Oliver's (Oliver and Webster 1987) initial survey of the soil the Wyre Forest in England, the sampling with even coverage was too sparse; the distances between neighbouring sampling points exceeded the range of spatial correlation in the soil variables.

We return to our search for that range below. Before that we state some general principles.

1. The maximum lag to which you compute the variogram should exceed the correlation range, and if it exists the sampling plan should ensure that.
2. The steps by which the lag is incremented should be small enough and the number of lags large enough for the experimental estimates to reveal the functional form of the variogram. Ideally you should aim for about six estimates within the correlation range, if it exists, and another four beyond, and sampling should be designed to provide them.
3. The size of sample should be large enough to place the estimates of the semi-variances within acceptable confidence limits. A good working rule is to aim for at least 100–150 sampling points.

You might be able to judge the first from your understanding of the environment and from visible features of the landscape; physiography is a good guide. Alter-natively, or in addition, you might already have or know of empirical variograms for similar land nearby. Item 2 depends to some extent on item 1, because only if you know the correlation range can you decide the interval between estimates and the sampling intervals on the ground to provide them. If you want the variogram solely for kriging then you should have one that is well estimated at short lag distances, and you should design a scheme that includes many pairs of points separated by short distances. Therefore, for a grid survey sample more intensively from randomly selected nodes to provide such pairs of points (Fig. 5.5).

Item 3 is widely misunderstood. You cannot apply the classical formula based on χ^2 to obtain confidence intervals on the experimental variogram calculated by the method of moments, Eq. (3.1), because the same data are used many times over and successive estimates are correlated. The advice in several texts to aim for 30–50 pairs of comparisons in each estimate, $m(\mathbf{h})$ in Eq. (3.1), is seriously misleading. It implies fewer than 50 points for a grid in two dimensions, and we know from empirical studies (Webster and Oliver 1992) that it leads almost inevitably to poor estimates and to erratic variograms.

We have already drawn attention to this shortcoming in Chap. 3, and we reinforce the matter in Fig. 3.5. That figure shows confidence intervals on experimental variograms computed from samples of four sizes. The upper two, Fig. 3.5a, b, in the figure for samples of size 49 and 81 are wide at all lags. As we have stated before, you should aim to sample at 100–150 points to obtain a reliable variogram.

5.1.1 Nested Sampling

Surveyors often have little or no idea of the range of spatial dependence or of the form of the variogram within its range. This is especially true when they begin investigations in unfamiliar regions. Guesswork can be expensive, either because the sampling is too sparse resulting in a variogram that is all nugget and is useless for kriging or because it is unnecessarily dense. In these circumstances sampling can be staged, with the first stage one of nested sampling followed by hierarchical analysis of variance (ANOVA) or its equivalent by REML.

The aim of such a scheme is to estimate efficiently the contribution made to the variation over scales ranging widely from fine to coarse in the region. The general principle was first proposed by Youden and Mehlich (1937) for sampling soil. Although the authors' original paper lay buried for a long time the technique was rediscovered and is now well documented in texts by Webster and Oliver (2007) and Webster and Lark (2013). The latter includes several novel options in an attempt to optimize the approach. Here we concentrate on the basic features of the strategy.

Stages are defined in terms of spacings between sampling points. At the lowest stage pairs or triplets of points are separated by the shortest distance of interest. At the highest stage, stage 1, pairs or triplets of groups are separated by the largest distance of interest. In between are several stages with points separated by intermediate distances. The distances progress in geometric sequence such that at any stage above the lowest the distance is at least 3 times that of the one below. The separating distances are fixed, but the orientations of the separations are chosen at random. The effects of distance are assumed to be random, and so the appropriate model for the analysis of variance is Model II of Marcuse (1949).

For a design with p stages the model of variation is

$$Z_{ijk...m} = \mu + A_i + B_{ij} + C_{ijk} + \cdots + \varepsilon_{ijk}. \tag{5.1}$$

The quantity μ is the mean, and the A_i, B_{ij}, C_{ijk}, ..., $\varepsilon_{ijk...m}$ are independent random variables associated with stages 1, 2, 3, ..., p, respectively, with means of zero and variances $\sigma_1^2, \sigma_2^2, \sigma_3^2, \ldots, \sigma_p^2$. These latter are the components of variance for the p stages, and each one is a measure of the variation attributable to that stage, i.e. to that separating distance. Together they sum to the total variance:

$$\sigma^2 = \sigma_1^2 + \sigma_2^2 + \sigma_3^2 + \cdots + \sigma_p^2. \tag{5.2}$$

Miesch (1975) pointed out that if estimates of these components are accumulated, starting with that at the smallest spacing, they form a first approximation to the experimental variogram, thus:

$$
\begin{aligned}
\widehat{\sigma}_p^2 &= \widehat{\gamma}(h_p) \\
\widehat{\sigma}_{p-1}^2 + \widehat{\sigma}_p^2 &= \widehat{\gamma}(h_{p-1}) \\
\widehat{\sigma}_{p-2}^2 + \widehat{\sigma}_{p-1}^2 + \widehat{\sigma}_p^2 &= \widehat{\gamma}(h_{p-2}),
\end{aligned} \tag{5.3}
$$

and so on, where the h_p, h_{p-1}, h_{p-2}, ..., h_1 are separating distances equivalent to the lag distances in geostatistical convention.

The analysis of variance for Model II above can be set out as in Table 5.1 in which there are four stages and N data, each of which belongs to one and only one group in each stage.

The table is quite general. It can be extended for more than four stages, and it can be simplified for fully balanced designs in which the same number of divisions is made at any particular stage into groups at the stage below. Balanced designs are attractive statistically because they lead to a straightforward analysis, and the variance components are readily calculated from the table because, for example, $u_{3,3} = u_{2,3} = u_{1,3}$ and $u_{2,2} = u_{1,2}$. Their big disadvantage is that the number of sampling points increases exponentially, at least two-fold for each additional stage, as the number of stages increases and soon becomes unaffordable.

Balance is not necessary, however, because one does not need the very many degrees of freedom in the low stages to obtain reliable estimates of the components. Unbalanced designs can still be analysed by ANOVA, but calculating the components of variance is more complex because their coefficients, the u in the table, change from stage to stage. Gower (1962) devised formulae for calculating the coefficients, and a worked example appears in the 6th edition of *Statistical Methods* of Snedecor and Cochran (1967), but not in later editions. For theoretical reasons we now prefer to estimate the components by residual maximum likelihood (REML) as described by Webster et al. (2006).

For balanced designs the results are the same, but for unbalanced ones they generally differ somewhat.

Table 5.1 Hierarchical analysis of variance

Stage	Degrees of freedom	Parameters estimated by mean squares
Stage 1	$f_1 - 1$	$u_{1,1}\sigma_1^2 + u_{1,2}\sigma_2^2 + u_{1,3}\sigma_3^2 + \sigma_4^2$
Stage 2	$f_2 - f_1$	$u_{2,2}\sigma_2^2 + u_{2,3}\sigma_3^2 + \sigma_4^2$
Stage 3	$f_3 - f_2$	$u_{3,3}\sigma_3^2 + \sigma_4^2$
Residual (stage 4)	$N - f_3$	σ_4^2
Total	$N - 1$	

5.1.1.1 Illustrative Example: Nested Sampling in the Wyre Forest

Following the initial survey of the soil of the Wyre Forest Oliver (Oliver and Webster 1987) planned a second one to discover the scale(s) of variation in the soil. The sampling comprised nine principal nodes on a grid at intervals of 600 m; this was stage 1. The points for stage 2 were selected 190 m from each node in a random direction. From each point in stage 2 a point was selected 60 m away to form stage 3, and from each of those points another was chosen 19 m away (stage 4). Finally, from half of the stage 4 points, points were chosen 6 m away to form the fifth stage. This gave $9 \times 2 \times 2 \times 2 = 72$ sampling points in the first four stages plus a further 36 in the fifth stage, giving 108 points in all. The structure of the scheme is shown as a topological tree in Fig. 5.1. The hierarchy is unbalanced in that at stage 4 only half of the sampling points have pairs in stage 5. Figure 5.2 shows the sampling configuration on the ground for one node.

The design might not have been optimal, but it was almost certainly a better use of resources than a balanced design, and, perhaps surprisingly, better than a design that distributes the degrees of freedom equally among the stages (Webster and Lark 2013).

Oliver and Webster originally estimated the components of variance by Gower's method, but later they re-analysed their data by REML (Webster et al. 2006), Table 5.2 lists the resulting components for three depths.

Fig. 5.1 Topology of one branch of the nested sampling scheme by Oliver (see Oliver and Webster 1987) to sample the soil of the Wyre Forest. Notice that only half of the branches at Stage 4 (19 m) are divided in the unbalanced design

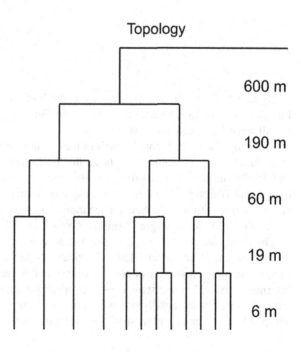

Topology

600 m

190 m

60 m

19 m

6 m

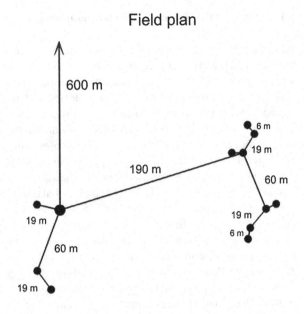

Fig. 5.2 Sampling plan of sites for one of the main branches from a grid node in the Wyre Forest with distances 190, 60, 19 and 6 m (Oliver and Webster 1987)

Table 5.2 Components of variance of percentage of sand in the soil of the Wyre Forest estimated by REML (from Webster et al. 2006)

Source (stage)	Distance/m	Components of variance		
		Depth/cm		
		0–5	25–30	50–55
1	600	38.12	16.68	33.75
2	190	−58.03	−90.02	−100.19
3	60	102.50	198.51	314.81
4	19	131.50	131.96	−38.89
5 (residual)	6	54.9	108.56	303.26

By accumulating the components from the bottom of the table upwards, as in Eq. (5.3), we obtain the variograms shown in Fig. 5.3. The variograms are erratic, but all three have maxima at 60 m.

Evidently, the range is roughly half of the distances between neighbouring points in the first survey. The figure also shows that for the first and second depths, 0–5 cm and 25–30 cm, a large proportion of the variance is between 6 and 60 m. Oliver (Oliver and Webster 1987) went on to sample the region at 5-m intervals on transects at various orientations and obtained accurate variograms by the method of moments and modelled them for kriging from data on a grid with nodes at 20-m intervals.

The above shows something of what can be achieved by splitting survey into distinct stages. Marchant and Lark (2006, 2007) developed this line of investigation, combining estimation of the variogram and kriging in stages such that the information gained in one stage is used to adapt the sampling in the next, and so on with the hope that eventually one would be able to predict and map a variable with acceptable confidence within specified budgets, starting, as it were, with a blank

Fig. 5.3 Approximate variograms of percentage sand at three depths from the nested survey of the Wyre Forest obtained by accumulating the components of variance estimated by REML

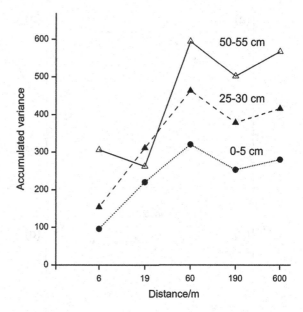

sheet of paper. We leave the reader to pursue their strategy in the papers mentioned and in the book chapter by Marchant and Lark (2010).

What should you do if you must do the field work in a single stage? This is often the case, perhaps because of logistic difficulties and costs of getting to remote regions, perhaps because clients want quick assessments, perhaps because money is available for only a single season in the field. In these situations surveyors find that they must sample in such a way as to estimate the variogram and model it and krige from the same set of data. They cannot expect to optimize any of the steps. Pragmatically, a surveyor must start somewhere. One starting point, mentioned already, is prior knowledge of the region, especially of the landscape and physiography if one is dealing with attributes of the soil or land more generally. That should enable one to decide sampling intervals on transects for estimating the variogram and perhaps wider ones on a grid for the kriging. One will not know what the maximum errors are until one has finished, and that is a hazard.

The example below shows how variograms of ancillary data from aerial photographs, sensors and yield monitors, and existing variograms of the properties of interest can be used to guide sampling for future surveys. The data are from a 23-ha field on the Yattendon Estate, Berkshire, England (Oliver and Carroll 2004). A colour aerial photograph for 1991 was digitized and the variogram computed from the digital numbers for the red waveband. Figure 5.4a shows the experimental variogram and the fitted nested spherical model, Eq. (3.12), and Table 5.3 lists the model parameters. The yield of wheat was recorded in the field in 1995 and the variogram was computed and modelled. Figure 5.4b shows the experimental values and the fitted nested spherical function, and Table 5.3 lists the parameters of that model. The topsoil (0–15 cm) was sampled on a 30-m grid with additional samples at

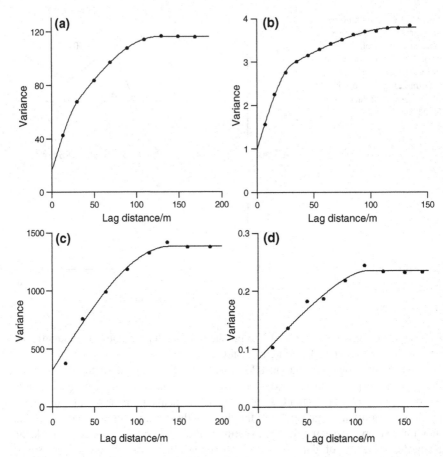

Fig. 5.4 Experimental variograms and fitted models of: **a** red waveband of a digitized colour aerial photograph taken in 1991, **b** wheat yield recorded in 1995, **c** potassium of the topsoil (0–15 cm) and **d** subsoil pH (30–60 cm) for a field on the Yattendon Estate, Berkshire, UK

Table 5.3 Model parameters of soil and ancillary data for the Yattendon Estate

Variable	Model type	Estimates of parameters				
		c_0	c_1	c_2	a_1/m	a_2/m
Soil						
Potassium—0–30 cm	Spherical	318.4	1065.0		140.1	
pH—30–60 cm	Circular	0.0824	0.152		109.8	
Ancillary						
Aerial image 1991—red waveband	Double spherical	16.86	24.91	74.52	32.66	126.8
Yield—1995	Double spherical	0.995	1.494	1.311	32.37	127.6

randomly selected grid nodes 15 m apart, and the subsoil (30–60 cm) was sampled on a 60-m grid with additional samples at selected grid nodes 15 m and 30 m apart. The experimental variogram and fitted spherical function, Eq. (3.10), of topsoil available potassium are shown in Fig. 5.4c, and the model parameters are listed in Table 5.3. Figure 5.4d shows the experimental variogram of subsoil pH with a circular function fitted, Eq. (4.16); the model parameters are listed in Table 5.3. Note that the variogram ranges of the longer structure for the aerial photograph and yield, and the ranges for potassium and pH are similar.

Kerry et al. (2010) suggested after repeated sampling of a large set of simulated values that sampling at 0.33 or less of the variogram range would provide an adequate basic grid. The average range of the variograms examined in the above example is about 126 m, and sampling at 0.33 times the range of the variogram would give an interval of 42 m for the grid. However, we recommend strongly that additional samples are taken at intervening intervals as above for the field at Yattendon to ensure that the variogram is estimated well near to the origin.

Aerial photographs are an excellent source of information for environmental surveys where the patterns of variation they show are linked with those of the variables of concern. Variograms can be computed from the digitized values prior to field work and used to guide the sampling. Milne et al. (2010) made good use of them in their analysis of gilgai patterns in Australia.

An alternative starting point is the budget; that will determine the total number of sampling points. If all the points are placed on a grid then the interval might be too large to estimate the variogram; there might be no comparisons from which to estimate the semivariances at short enough lags.

Atteia et al. (1994) planned their survey, which had to be done in a single season, with random nested sampling around 23 of their grid nodes. More often practitioners place their additional sampling points on some of the grid lines joining the nodes, as in Fig. 5.5. In Fig. 5.5a the additional points are 1.1 and 1.3 units

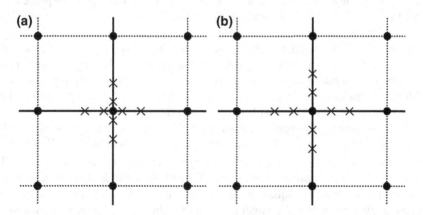

Fig. 5.5 Configurations for additional sampling at the node of a square sampling grid. The supplementary sampling points are shown as *crosses* at distances of: **a** 0.1 and 0.3 times the grid interval from the central node marked by a *circle* and **b** 0.2 and 0.4 times the interval

away from the central node. This would allow one to compute semivariances, $\hat{\gamma}(h)$, at lag distances 0.1, 0.2, 0.3, 0.4, 0.6, 0.7 and 0.9 units on the principal axes. In Fig. 5.5b the additional points are placed 0.2 and 0.4 units away from the central node, and that allows one to compute $\hat{\gamma}(h)$ at lag distances of 0.2, 0.4, 0.6 and 0.8 units. These schemes are not optimal, but both are better than a strict grid in that they enable one to compute and model the variogram over lags distances shorter than the grid interval and which one needs for predicting values between the nodes.

5.2 Sampling Plans for Mapping

The prediction of variables at unvisited places without bias is a central aim of geostatistics, and in Chap. 4 we presented the kriging equations to achieve that. The kriging equations also minimize the variance of any prediction, and their solution leads to an estimate of the kriging variance or error. In addition to being able to map a variable at a fine resolution from sample data we can also map the kriging variance or its square root, the kriging error. Such a map might show where extra sampling is needed to diminish the error and increase confidence. We can also use the kriging equations to plan sampling to map within some tolerable error—provided we have an accurate model of the variogram.

You can see that Eqs. (4.3) and (4.5) contain only semivariances, which derive from the variogram and the configuration of the sampling points in relation to the target point or block. They do not depend on the observed values at the sampling points. If you know the variogram then you can add points to the kriging systems where data seem to be too sparse and calculate what the kriging variances would be if you sampled at those points. To some extent choosing the additional sampling points is a matter of trial and error. You add a point where the existing kriging variance is greatest and solve the new kriging system, and you repeat the procedure until the kriging error is small enough everywhere.

If you know the variogram beforehand you can plan a sampling that is nearly optimal in that it will minimize the maximum kriging variance for a given cost. In general, the further a target point is from data the larger is the kriging variance. You can minimize the maximum distance between target and data by sampling on a grid; in those circumstances the maximum distance is from the centre of a grid cell to the nearest grid nodes. For punctual kriging the kriging variance is greatest there.

These maximum distances are minimized for a given sampling density with triangular configurations, and the maximum kriging variance is least. Figure 5.6 shows the situation. For a square grid the maximum distance is $1/\sqrt{2} \approx 0.7071$ units, whereas for an equilateral triangular grid with the same density the maximum distance is 0.6204 units. Square grids are more convenient, however, and as the maximum distance between a target point and the distance to the nearest sampling

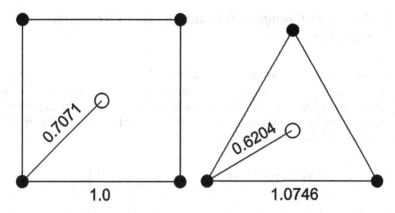

Fig. 5.6 Distances between the centres of grid cells and nearest sampling points for square and equilateral triangular grids with the same sampling density of one point per unit area

points is little more than for triangular grids of the same density and there are four near points instead of three the maximum kriging variance is only slightly larger: see Fig. 8.23a in Webster and Oliver (2007) and Fig. 9.7a in Webster and Lark (2013).

Note, however, that kriging variances tend to increase as the margins of the region are approached and that for irregularly shaped regions a regular grid should be modified to achieve best results.

The following procedure, proposed by Burgess et al. (1981) and reiterated by Webster and Lark (2013), will enable you to plan a grid.

1. Set up the kriging equations for a square configuration of sampling points with the target point or block at its centre.
2. Solve the equations for a small sampling interval, the smallest that is likely to be feasible, and compute the kriging variance.
3. Increase the sampling interval in steps and repeat the calculations in 2 above at each step.
4. Draw a graph of kriging variance (or its square root, the kriging error) against the sampling interval and link the points by a smooth curve.
5. Draw a horizontal line on this graph at your chosen maximum variance or error to cut the curve, and drop a perpendicular from the intersection to the abscissa.

That perpendicular gives the required sampling interval, from which you can determine the number of sampling points for mapping and hence the budget. Alternatively, if the budget for survey is fixed then that will determine the sampling interval, and you follow step 5 in reverse. You draw a perpendicular from the abscissa to cut the curve and read the corresponding maximum kriging variance or error on the ordinate.

5.2.1 Illustrative Example: Sampling to Map Chromium in the Swiss Jura

Atteia et al. (1994) sampled the topsoil of part of the Swiss Jura in a survey of potential toxicity caused by heavy metals, among which was included chromium (Cr). From 366 measurements they obtained the omnidirectional experimental variogram shown by the points plotted in Fig. 5.7 and to which they fitted an exponential model with equation

$$\gamma(h) = 19.98 + 98.34 \times \left\{ 1 - \exp\left(-\frac{h}{174} \right) \right\}. \qquad (5.4)$$

Here h is the lag distance, and the distance parameter, $a = 174$, is in metres.

Using this variogram we can calculate the maximum kriging variances or errors for points or blocks of any reasonable size against sample spacing by following steps 1 to 4 above. Usually we shall be interested in blocks, and in Fig. 5.8 we show the maximum kriging errors for two sizes of block, 50 m × 50 m (= 0.25 ha) and 100 m × 100 m (= 1 ha), as the curves.

If the grid interval is much shorter than the side of the block the maximum kriging variance can occur for blocks centred on grid nodes (Burgess et al. 1981; Webster and Lark 2013), but the differences between it and that from cell-centred blocks are small and of little practical significance.

Let us suppose that in some future survey the maximum kriging error is to be no more than 10 % of the tolerable maximum concentration. The threshold for Cr set in the VSBo of 1986 for Switzerland (FOEFL 1987) is 75 mg kg^{-1} of soil. That leads

Fig. 5.7 Variogram of chromium in the topsoil in the Swiss Jura. The *line* is the fitted exponential model with the parameters shown on the graph

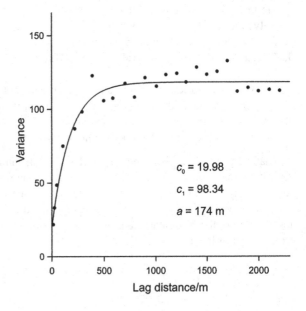

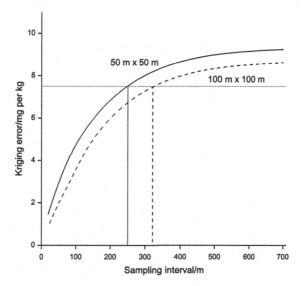

Fig. 5.8 Maximum kriging errors for chromium in the topsoil in the Swiss Jura for square blocks of 0.25 and 1 ha. The *horizontal line* is drawn at concentration 7.5 mg kg^{-1}, which is 10 % of the tolerable maximum set in the VSBo (FOEFL 1987)

to a maximum tolerable kriging error of 7.5 mg kg^{-1}. So we draw a horizontal line at that value to cut the curves and drop the perpendiculars shown in the figure. For 0.25-ha blocks the spacing is 245 m, and for the 1-ha blocks it is 322 m.

5.3 Summary

We can provide guidelines for sampling for geostatistical interpolation and mapping *if you have a satisfactory model for the variogram*. The best advice is to sample on a grid, for which either the survey budget or the maximum tolerance on a prediction determines the grid interval. If you have to estimate the variogram first and have little idea of its form then the best approach is to survey in stages, beginning with a nested scheme with analysis by REML to estimate the spatial components of variance, followed by systematic sampling to estimate the variogram and model it, and finally a grid for the mapping. If the survey cannot be staged then your best approach is to survey on a grid with its interval determined by whatever information you can glean from existing sources and an understanding of the landscape—or by the budget if that is fixed—and augment the grid with additional sampling points between the grid nodes.

Chapter 6
Dealing with Trend

Abstract Where there is trend, i.e. smooth variation in space, also known as drift, the experimental variogram of the observations is no longer a function solely of a random variable. The assumption of stationarity no longer holds. An experimental variogram that increases with ever increasing gradient as the lag distance increases is usually symptomatic of trend. In these circumstances the process should be modelled as a combination of a deterministic trend plus spatially correlated random residuals from the trend. Estimation of the trend by ordinary least squares regression and a separate analysis of the residuals lead to bias in the variogram. Best practice is to estimate the trend and the parameters of the variogram by residual maximum likelihood (REML). Once this has been achieved, one can use Matheron's universal kriging for prediction. The technique embodies simple functions of the coordinates that take the trend of given order into account. The methods are illustrated with an example of trend in the soil's sand content of a field.

Keywords Trend · Deterministic variation · Ordinary least squares · Residual maximum likelihood (REML) · Universal kriging · Empirical best linear unbiased predictor (E-BLUP)

6.1 Trend

The assumption of intrinsic stationarity, which underlies the analyses in Chaps. 2 and 3 and which has seemed easily satisfied, is not satisfactory in two situations. One is where there is pronounced geographic trend, known to geostatisticians as 'drift'.

What is regarded as trend depends to some extent on the scale at which the variation is viewed, but the basic idea is that trend is smooth systematic non-random variation. It might be regional, i.e. systematic variation over the whole region of interest, or local from point to point within the region. You might detect the first in one dimension by plotting the data from a transect as in Fig. 6.1a and

© The Author(s) 2015   85
M.A. Oliver and R. Webster, *Basic Steps in Geostatistics: The Variogram and Kriging*, SpringerBriefs in Agriculture, DOI 10.1007/978-3-319-15865-5_6

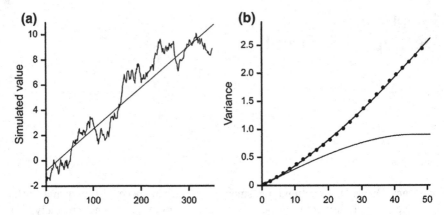

Fig. 6.1 a Simulated sequence of 350 values with a linear trend plus realization from a correlated random process superimposed. The *straight line* is that of the trend estimated by REML. **b** Variograms of the trace. The *black discs* are the experimental semivariances estimated by the method of moments with a power model fitted; the *lower line* is the spherical model fitted to the residuals from the trend estimated by REML

in two dimensions on a 'posting' with symbols of size proportional to the recorded values. Alternatively, you might use a standard interpolation routine in a graphics package to produce an isarithmic ('contour') map that might reveal the presence of long-range trend. Realize, however, that the interpolated values are not necessarily sound statistically. Trend also manifests itself in experimental variograms of data. Almost always it appears as ever-increasing semivariances with increasing lag distance. The experimental variogram of the data in Fig. 6.1a, shown by the black discs in Fig. 6.1b, increases without bound and is a symptom of a regional trend. Such a trend violates the assumptions of stationarity, and we must express the variation in a way that includes the trend. The solution to the problem posed in this particular instance is explained below.

This kind of change can be described by a deterministic function such as a trend surface, and values can be predicted by a mathematical function.

Recall from Chap. 2 that in the stationary case we can express the variation simply as

$$Z(\mathbf{x}) = u(\mathbf{x}) + \varepsilon(\mathbf{x}), \qquad (6.1)$$

in which $u(\mathbf{x}) = \mu$ is the mean of the process, a constant for all $\mathbf{x}$, and $\varepsilon(\mathbf{x})$ is a spatially correlated random residual. If there is trend then $u(\mathbf{x})$ is no longer constant but depends on $\mathbf{x}$. Further, the experimental variogram of the data, $z(\mathbf{x}_i)$, $i = 1, 2, \ldots$, no longer estimates the variogram of the random residuals, $\varepsilon(\mathbf{x})$, as it does in the stationary case. Instead we want an estimate of the variogram of $\varepsilon(\mathbf{x}) = Z(\mathbf{x}) - u(\mathbf{x})$; that is the variogram needed for kriging.

If we have a function, $\gamma(\mathbf{h})$, for $\varepsilon(\mathbf{x})$ then we can predict Z at any $\mathbf{x}_0$ by what Matheron (1969) called 'universal kriging'. Such a prediction is still a linear sum:

$$\hat{Z}(\mathbf{x}_0) = \sum_{i=1}^{N} \lambda_i f_k(\mathbf{x}_i), \tag{6.2}$$

with weights λ_i, $i = 1, 2, \ldots, N$. In addition are the f_k. These are simple functions of the spatial coordinates, which are best referred to the target point $\mathbf{x}_0$. For a linear trend there are three ($K + 1 = 3$), with values

$$f_0 = 1, \quad f_1 = x_1, \quad \text{and} \quad f_2 = x_2.$$

For quadratic trend there are three additional functions:

$$f_3 = x_1^2, \quad f_4 = x_1 x_2, \quad \text{and} \quad f_5 = x_2^2,$$

giving $K + 1 = 6$ in all. These are incorporated in the universal kriging system, which is no more than an augmentation of the ordinary kriging system (see Sect. 4.2), thus:

$$\mathbf{A} = \begin{bmatrix}
\gamma(\mathbf{x}_1,\mathbf{x}_1) & \gamma(\mathbf{x}_1,\mathbf{x}_2) & \cdots & \gamma(\mathbf{x}_1,\mathbf{x}_N) & 1 & f_1(\mathbf{x}_1) & f_2(\mathbf{x}_1) & \cdots & f_K(\mathbf{x}_1) \\
\gamma(\mathbf{x}_2,\mathbf{x}_1) & \gamma(\mathbf{x}_2,\mathbf{x}_2) & \cdots & \gamma(\mathbf{x}_2,\mathbf{x}_N) & 1 & f_1(\mathbf{x}_2) & f_2(\mathbf{x}_2) & \cdots & f_K(\mathbf{x}_2) \\
\cdot & \cdot & \cdots & \cdot & \cdot & \cdot & \cdot & \cdots & \cdot \\
\cdot & \cdot & \cdots & \cdot & \cdot & \cdot & \cdot & \cdots & \cdot \\
\cdot & \cdot & \cdots & \cdot & \cdot & \cdot & \cdot & \cdots & \cdot \\
\gamma(\mathbf{x}_N,\mathbf{x}_1) & \gamma(\mathbf{x}_N,\mathbf{x}_2) & \cdots & \gamma(\mathbf{x}_N,\mathbf{x}_N) & 1 & f_1(\mathbf{x}_N) & f_2(\mathbf{x}_N) & \cdots & f_K(\mathbf{x}_N) \\
1 & 1 & \cdots & 1 & 0 & 0 & 0 & \cdots & 0 \\
f_1(\mathbf{x}_1) & f_1(\mathbf{x}_2) & \cdots & f_1(\mathbf{x}_N) & 0 & 0 & 0 & \cdots & 0 \\
f_2(\mathbf{x}_1) & f_2(\mathbf{x}_2) & \cdots & f_2(\mathbf{x}_1) & 0 & 0 & 0 & \cdots & 0 \\
\cdot & \cdot & \cdots & \cdot & \cdot & \cdot & \cdot & \cdots & \cdot \\
\cdot & \cdot & \cdots & \cdot & \cdot & \cdot & \cdot & \cdots & \cdot \\
f_K(\mathbf{x}_1) & f_K(\mathbf{x}_2) & \cdots & f_K(\mathbf{x}_N) & 0 & 0 & 0 & \cdots & 0
\end{bmatrix}$$

$$\boldsymbol{\lambda} = \begin{bmatrix} \lambda_1 \\ \lambda_2 \\ \cdot \\ \cdot \\ \cdot \\ \lambda_N \\ \psi_0 \\ \psi_1 \\ \psi_2 \\ \cdot \\ \cdot \\ \psi_K \end{bmatrix} \quad \text{and} \quad \mathbf{b} = \begin{bmatrix} \gamma(\mathbf{x}_1,\mathbf{x}_0) \\ \gamma(\mathbf{x}_2,\mathbf{x}_0) \\ \cdot \\ \cdot \\ \cdot \\ \gamma(\mathbf{x}_N,\mathbf{x}_0) \\ 1 \\ f_1(\mathbf{x}_0) \\ f_2(\mathbf{x}_0) \\ \cdot \\ \cdot \\ f_K(\mathbf{x}_0) \end{bmatrix},$$

or in matrix notation

$$\mathbf{A}\lambda = \mathbf{b}. \tag{6.3}$$

Notice that there are now three Lagrange multipliers, ψ_0, ψ_1 and ψ_2, for a linear trend and three more, ψ_3, ψ_4 and ψ_5, for a quadratic trend. As in ordinary kriging, matrix $\mathbf{A}$ is inverted, and the weights and the Lagrange multipliers are obtained as

$$\lambda = \mathbf{A}^{-1}\mathbf{b}. \tag{6.4}$$

The weights are inserted into Eq. (6.2), and the kriging variance is given by

$$\sigma_{UK}^2 = \mathbf{b}^{T}\lambda. \tag{6.5}$$

Also as in ordinary kriging, we can usually work within a window with many fewer data than the whole set of size N, and once we have decided the size of the window the procedure is automatic.

6.1.1 Variogram and Model

Although the universal kriging system appears as a simple augmentation of the system for ordinary kriging, the semivariances in matrix $\mathbf{A}$ and vector $\mathbf{b}$ are not those of variable Z itself; instead, they are the semivariances of the residuals from the trend as we indicated above. To estimate them we must separate the trend, a deterministic term, from the residuals, which we treat as random. Matheron did not say how this might be done, and for some three decades there was no entirely satisfactory way of estimating without bias the two components of the model or of combining them for prediction.

The breakthrough came when Stein (1999) pointed out that the several kinds of kriging then current are all forms of an empirical best linear unbiased predictor, E-BLUP, (universal kriging is one) and that the two components of the model could be estimated simultaneously by likelihood methods. We now know that the result is best achieved by residual maximum likelihood (REML) (Lark et al. 2006; Minasny and McBratney 2007). For this we re-formulate Eq. (6.1) as

$$Z(\mathbf{x}) = \mathbf{w}\beta + \varepsilon(\mathbf{x}), \tag{6.6}$$

in which the vector $\mathbf{w}$, with $K + 1$ columns, contains the elements $1, x_1, \ldots, x_K$ of the trend function and vector β contains the coefficients.

We can represent the data similarly:

$$\mathbf{z}(\mathbf{X_d}) = \mathbf{W_d}\boldsymbol{\beta} + \boldsymbol{\varepsilon}(\mathbf{X_d}). \qquad (6.7)$$

For N data, matrix $\mathbf{X_d}$ with N rows and two columns for the two spatial coordinates contains the positions of the data (denoted by subscript d). The vectors $\mathbf{z}$ and $\boldsymbol{\varepsilon}$ also have N rows. Vector $\mathbf{w}$ of Eq. (6.6) is replaced by $\mathbf{W_d}$, known as a 'design matrix', with N rows and $K + 1$ columns. We assume that the random components are second-order stationary and jointly normally distributed with means of zero and a covariance matrix $\mathbf{C_{dd}}$. This matrix is obtained from the covariance function, $C(\mathbf{h})$, which because the process is assumed to be second-order stationary is equivalent to the variogram of the residuals:

$$\gamma(\mathbf{h}) = C(\mathbf{0}) - C(\mathbf{h}), \qquad (6.8)$$

where $C(\mathbf{0}) = \sigma^2$ is the variance of the random process.

The variogram is necessarily bounded, and as in Chap. 2, it can usually be described by a simple function with three parameters, namely a nugget variance c_0, a sill of the correlated structure c and a distance parameter a. These parameters, which are of the random component $\varepsilon(\mathbf{x})$ in Eq. (6.1), must be estimated from data; and for this the random component must be separated from the trend. Otherwise the parameters would be biased because they would depend on $\boldsymbol{\beta}$. The solution to the problem is to maximize the log-likelihood of the residuals, given the data: $L[c_0, c, a \,|\mathbf{z}(\mathbf{X_d})]$. The values of c_0, c and a that maximize L are found numerically, and with those one can obtain estimates of the fixed effects of $\boldsymbol{\beta}$ by generalized least squares approximation.

You can find a full account of the solution in Webster and Oliver (2007).

6.2 Example

As an example we examine the data displayed in Fig. 6.1a. Evidently the sequence of values has a strong linear trend. Table 6.1 summarizes the statistics. The sequence's experimental variogram computed by the method of moments appears as the black discs in Fig. 6.1b. It increases with an ever increasing gradient, and it can fitted by a power function, which is shown by the line passing through the points. The exponent of the model is 1.29 (Table 6.2), and it lies within the tolerable range, i.e. $0 < \alpha < 2$, for a purely random process, yet the process is clearly far from purely random. We cannot accept that variogram as a description of the process; we must separate the random process from the trend.

Given that the trend is so obviously linear, we can propose the following model for the whole one-dimensional process:

Table 6.1 Statistical summary of simulated transect

Number of observations	350
Minimum	−1.575
Maximum	10.09
Mean	5.117
Median	6.456
Variance	12.802
Standard deviation	3.578
Skewness	−0.304

Table 6.2 Parameters of the power function fitted to the experimental variogram of the simulated trace in Fig. 6.1a

Model parameters	
Nugget, c_0	0.0283
Gradient, g	0.01641
Exponent, α	1.288

The model is given by $\gamma(h) = c_0 + gh^\alpha$

Table 6.3 Parameters of linear trend of the simulated trace in Fig. 6.1a and of a spherical variogram of residuals from the trend estimated by REML

Trend		Variogram		
Mean	Gradient	c_0	c	R
5.013	0.03283	0.0004	0.9016	45.9

The model is given by

$$\gamma(h) = \begin{cases} c_0 + c\left\{\frac{3h}{2r} - \frac{1}{2}\left(\frac{h}{r}\right)^3\right\} & \text{for } 0 < h \le r \\ c_0 + c & \text{for } h > r \\ 0 & \text{for } h = 0, \end{cases}$$

$$Z(x) = \mu + x\beta + \varepsilon(x), \tag{6.9}$$

in which μ is the mean of the process, β is a coefficient, and $\varepsilon(x)$ is a correlated random residual with covariance function $C(h)$ and variogram $\gamma(h)$. We can estimate the parameters from the data by REML. By assuming that the variogram belongs to the spherical family, we obtain the estimates listed in Table 6.3. The spherical component appears in Fig. 6.1b as the curve lying well below the experimental variogram of the raw data.

6.3 Illustration from a Case Study

We illustrate the procedure with data from a case study in a field on the Yattendon estate in southern England where there is a significant trend in the sand content of the topsoil. The field was sampled at 30-m intervals on a square grid. Topsoil (0–15 cm) was taken with a 3-cm-diameter auger from rectangles of 5 m × 2 m and mixed to form bulked or composite samples. The sand content was measured

Table 6.4 Statistical summary of data on the sand content (%) in the topsoil at Yattendon	Number of observations	230
	Minimum	14
	Maximum	83
	Mean	50.8
	Median	51
	Variance	207.41
	Standard deviation	14.4
	Skewness	0.02

on each bulked sample by laser diffraction. Additional observations were made likewise at 15-m intervals along several short transects from randomly selected grid nodes. Table 6.4 summarizes these data. The experimental variogram of the raw data appear as the grey triangles in Fig. 6.3. The semivariances increase with increasing lag distance, apparently without bound. Their behaviour is characteristic of global trend in the data. They are not those of the random term $\varepsilon(\mathbf{x})$ in Eq. (6.1), to which we might fit a model for kriging.

One can remove the trend by the once-popular trend surface analysis, which is simply ordinary least-squares (OLS) regression of the measured values on the spatial coordinates. In this instance we have done so by fitting a quadratic trend surface, which accounted for approximately 46 % of the variance. Figure 6.2 displays the data in pixel form, in which the quadratic trend is evident from the small values in both the NE and SE of the map.

The experimental variogram of the OLS residuals is shown in Fig. 6.3 by the black discs. We fitted a spherical model to the experimental values, the dotted line, and the model parameters are listed in Table 6.2. The variogram is clearly bounded:

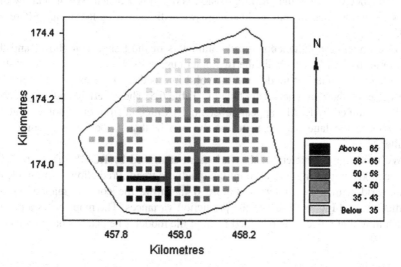

Fig. 6.2 Pixel map of the sand content, as percentage by weight, of the topsoil at Yattendon

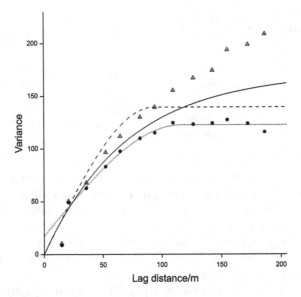

Fig. 6.3 Variograms of the sand content of the topsoil at Yattendon. The *grey triangles* are the experimental semivariances of the measurements. The *black discs* are the experimental semivariances of the residuals from the ordinary least squares quadratic trend surface with the spherical model fitted to them by weighted least squares approximation. The *solid line* is the exponential variogram of the residuals from the quadratic trend estimated by REML, and the *dashed line* is the analogous spherical variogram

the residuals can reasonably be regarded as the outcome of a second-order stationary process. They are, however, correlated, and therefore the assumption of independence on which the OLS regression is based is violated. We also know that this variogram is biased, and the bias increases with increasing lag distance (Cressie 1993).

By estimating simultaneously the parameters of the trend, in β above, and the variogram by REML we obtain the two other curves shown in Fig. 6.3. The dashed curve is of the spherical model and the solid line is that of the exponential model. The values of their parameters are listed in Table 6.5. The spherical and exponential models obtained by REML appear substantially different from one another, but we may choose the latter as the better of the two in that its deviance is somewhat smaller.

We used the parameters of the variogram estimated by REML (Table 6.5) for universal kriging taking into account the quadratic trend. Predictions were made on a fine grid with a 5-m interval. Figure 6.4a shows the contour map of kriged predictions and Fig. 6.4b that of the prediction variances. The map shows a patchy distribution that reflects the bounded form of the model. The patches have a random

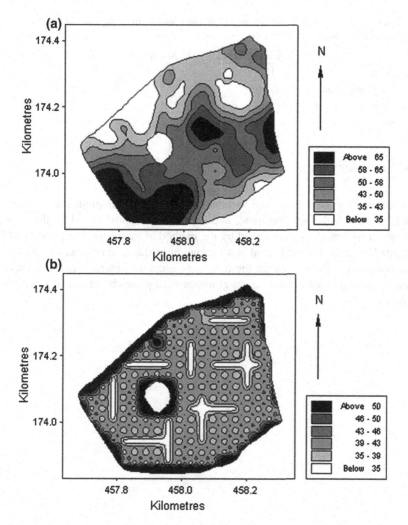

Fig. 6.4 Maps of **a** punctually kriged predictions and **b** punctually kriged predictions variances of sand content at Yattendon made by universal kriging and taking into account the quadratic trend

extent, which one would expect from an exponential function with its asymptotic upper bound. The white linear patches in the map of kriging variances represent the short transects along which sampling was more intense. The kriging variances within the circles are also small; these are in the vicinity of the sampling points. The largest kriging variances are at the edges of the field and around a small copse in the centre of the field that has been excluded from the map. The errors are large in these places because there were few samples from which to predict (Fig. 6.4b).

Table 6.5 Model parameters for variograms of the residuals from the quadratic trend of sand content (%) in the topsoil at Yattendon

	OLS spherical	REML spherical	REML exponential
Nugget, c_0	17.1	0	0
Structured component, c_1	105.6	139.4	171.5
Distance parameter (m)	116.8	93.4	70.9
Deviance		1261.2	1255.4

6.4 Summary

Where there is evident trend in a variable of interest the variogram is by definition that of the residuals from the trend, and it cannot be approximated by the experimental variogram of data computed by the method of moments. The trend must be separated from the residuals so that a variogram of the residuals can be estimated. Best practice is now to separate the two by residual maximum likelihood. Having done so, one can then predict values at unobserved places by universal kriging for mapping.

References

Adams, C. F., Harris, B. P., Marino, M. C., & Stokesbury, K. D. E. (2010). Quantifying sea scallop bed diameter on Georges Bank with geostatistics. *Fisheries Research, 106*, 460–467.

Araújo e Silva Ferraz, G., da Silva, F. M., de Carvalho Alves, M. de Lima Bueno, R., & da Costa, P. A. N. (2012). Geostatistical analysis of fruit yield and detachment force in coffee. *Precision Agriculture, 13*, 76–89.

Atteia, O., Webster, R., & Dubois, J.-P. (1994). Geostatistical analysis of soil contamination in the Swiss Jura. *Environmental Pollution, 86*, 315–327.

Blanchin, R., & Chilès, J.-P. (1993). The Channel Tunnel: Geostatistical prediction of the geological conditions and its validation by the reality. *Mathematical Geology, 25*, 963–974.

Brus, D. J., & de Gruijter, J. J. (1994). Estimation of non-ergodic variograms and their sampling variance by design-based sampling strategies. *Mathematical Geology, 26*, 437–454.

Burgess, T. M., Webster, R., & McBratney, A. B. (1981). Optimal interpolation and isarithmic mapping of soil properties. IV. Sampling strategy. *Journal of Soil Science, 32*, 643–654.

Bush, D. (2010). An overview of the estimation of kimberlite diamond deposits. In *The Southern African Institute of Mining and Metallurgy—Source to Use 2010* (pp. 73–84). Johannesburg: The Southern African Institute of Mining and Metallurgy.

Castrignanò, A., Boccaccio, L., Cohen, Y., Nestel, D., Kounatidis, I., Papadopoulos, N. T., De Bendetto, D., & Mavragani-Tsipidou, P. (2012). Spatio-temporal population dynamics and area-wide delineation of *Bactrocera oleae* monitoring zones using multi-variate geostatistics. *Precision Agriculture, 13*, 442–456.

Chilès, J.-P., & Delfiner, P. (1999). *Geostatistics: Modeling spatial uncertainty*. New York: Wiley.

Chilès, J.-P., & Delfiner, P. (2012). *Geostatistics: Modeling spatial uncertainty* (2nd ed.). New York: Wiley.

Cochran, W. G. (1977). *Sampling techniques* (3rd ed.). New York: Wiley.

Cressie, N. (1985). Fitting variogram models by weighted least squares. *Journal of the International Association of Mathematical Geology, 17*, 563–586.

Cressie, N. (1990). The origins of kriging. *Mathematical Geology, 22*, 239–252.

Cressie, N. A. C. (1993). *Statistics for spatial data*, revised edition. New York: Wiley.

Cressie, N. (2006). Block kriging for lognormal spatial processes. *Mathematical Geology, 38*, 413–443.

Cressie, N., & Hawkins, D. M. (1980). Robust estimation of the variogram. *Journal of the International Association of Mathematical Geology, 12*, 115–125.

De Gruijter, J. J., Brus, D. J., Bierkens, M. F. P., & Knotters, M. (2006). *Sampling for natural resources*. Berlin: Springer-Verlag.

Dowd, P. A. (1984). The variogram and kriging: Robust and resistant estimators. In G. Verly, M. David, A. G. Journel & A. Marechal (Eds.), *Geostatistics for natural resources characterization* (pp. 91–106). Dordrecht: D. Reidel.

© The Author(s) 2015
M.A. Oliver and R. Webster, *Basic Steps in Geostatistics: The Variogram and Kriging*, SpringerBriefs in Agriculture, DOI 10.1007/978-3-319-15865-5

Dubois, J.-P., Benitez, N., Liebig, T., Baudraz, M., & Okopnik, F. (2002). Le cadmium dans les sols du haut Jura Suisse. In D. Baize & M. Tercé (Eds.), *Les éléments traces métalliques dans les sols. Approches functionelles et spatiales* (pp. 33–52). Paris: INRA Editions.

Evans, K., Webster, R., Halford, P. D., Barker, A. D., & Russell, M. D. (2002). Site-specific management of nematodes—pitfalls and practicalities. *Journal of Nematology, 34*, 194–199.

Evans, K., Webster, R., Barker, A., Halford, P., Russell, M., Stafford, J., & Griffin, S. (2003). Mapping infestations of potato cyst nematodes and the potential for spatially varying application of nematicides. *Precision Agriculture, 4*, 149–162.

FOEFL (Swiss Federal Office of Environment, Forest and landscape) (1987). *Commentary on the ordinance relating to pollutants in soil.* Bern: FOEFL.

Gaus, I., Kinniburgh, D. G., Talbot, J. C., & Webster, R. (2003). Geostatistical analysis of arsenic concentration in groundwater in Bangladesh using disjunctive kriging. *Environmental Geology, 44*, 939–948.

Ge, Y., Thomasson, J. A., Sui, R. M., Morgan, C. L., Searcy, S. W., & Parnell, C. B. (2008). Spatial variation of fiber quality and associated loan rate in a dryland cotton field. *Precision Agriculture, 9*, 181–194.

Genton, M. G. (1998). Highly robust variogram estimation. *Mathematical Geology, 30*, 213–221.

Gower, J. C. (1962). Variance component estimation for unbalanced hierarchical classification. *Biometrics, 18*, 168–182.

Hbirkou, C., Welp, G., Rehbein, K., Hillnhütter, C., Daub, M., Oliver, M. A., & Pätzold, S. (2011). The effect of soil heterogeneity on the spatial distribution of *Heterodera schachtii* within sugar beet fields. *Applied Soil Ecology, 51*, 25–34.

Jardim, M., & Ribeiro, P. J. (2008). Geostatistical tools for assessing sampling designs applied to a Portuguese bottom trawl survey field experience. *Scientia Marina, 72*, 623–630.

Journel, A. G., & Huijbregts, C. J. (1978). *Mining geostatistics.* London: Academic Press.

Kerry, R., & Oliver, M. A. (2007a). Determining the effect of asymmetric data on the variogram. I. Underlying asymmetry. *Computers & Geosciences, 33*, 1212–1232.

Kerry, R., & Oliver, M. A. (2007b). Determining the effect of asymmetric data on the variogram. II. Outliers. *Computers & Geosciences, 33*, 1233–1260.

Kerry, R., Oliver, M. A., & Frogbrook, Z. L. (2010). Sampling in precision agriculture. In M. A. Oliver (Ed.), *Geostatistical applications for precision agriculture* (pp. 35–63). Dordrecht: Springer.

Kolmogorov, A. N. (1939). Sur l'interpolation et extrapolation des suites stationnaires. *Comptes Rendus de l'Académie des Sciences de Paris, 208*, 2043–2045.

Kolmogorov, A. N. (1941). Interpolirovanie I ekstapolirovanie stationarnykh sluchainykh posledovatel' nostei (Interpolated and extrapolated stationary random sequences). *Isvestia AN SSSR, Seriya Matematicheskaya, 5*(1).

Komnitsas, K., & Modis, K. (2009). Geostatistical risk estimation at waste disposal sites in the presence of hot spots. *Journal of Hazardous Materials, 164*, 1185–1190.

Krige, D. G. (1951). A statistical approach to some basic mine problems on the Witwatersrand. *Journal of the Chemical and Metallurgical Society of South Africa, 52*, 119–139.

Lark, R. M. (2000). A comparison of some robust estimators of the variogram for use in soil survey. *European Journal of Soil Science, 51*, 137–157.

Lark, R. M. (2002). Optimized spatial sampling of soil for estimation of the variogram by maximum likelihood. *Geoderma, 105*, 49–80.

Lark, R. M., Cullis, B. R., & Welham, S. J. (2006). On optimal prediction of soil properties in the presence of trend: The empirical best linear unbiased predictor (E-BLUP) with REML. *European Journal of Soil Science, 57*, 787–799.

Li, H. Y., Webster, R., & Shi, Z. (2015). Mapping soil salinity in the Yangtze delta: REML and universal kriging (E-BLUP) revisited. *Geoderma, 237–238*, 71–77.

Marchant, B. P., & Lark, R. M. (2006). Adaptive sampling and reconnaissance surveys for geostatistical mapping of the soil. *European Journal of Soil Science, 57*, 831–845.

Marchant, B. P., & Lark, R. M. (2007). Optimized sample schemes for geostatistical surveys. *Mathematical Geology, 39*, 113–134.

Marchant, B. P., & Lark, R. M. (2010). Sampling in precision agriculture, optimal designs from uncertain models. In M. A. Oliver (Ed.) *Geostatistical applications for precision agriculture* (pp. 65–87). Dordrecht: Springer.

Marcuse, S. (1949). Optimum allocation and variance components in mixed sampling with applications to chemical analysis. *Biometrics, 5*, 189–206.

Matheron, G. (1963). *Les variables régionalisées et leur estimation*. Paris: Masson.

Matheron, G. (1965). Principles of geostatistics. *Economic Geology, 58*, 1246–1266.

Matheron, G. (1969). *Le krigeage universel*. Cahiers du Centre de Morphologie Mathématique, No 1. Fontainebleau: Ecole des Mines de Paris.

McBratney, A. B., & Webster, R. (1986). Choosing functions for semivariograms of soil properties and fitting them to sampling estimates. *Journal of Soil Science, 37*, 617–639.

McBratney, A. B., Webster, R., McLaren, R. G., & Spiers, R. B. (1982). Regional variation of extractable copper and cobalt in the topsoil of south-east Scotland. *Agronomie, 2*, 969–982.

Meerschman, E., Cockx, E., Islam, M. M., Meeuws, F., & Van Meirvenne, M. (2011). Geostatistical assessment of the impact of World War I on the spatial occurrence of soil heavy metals. *Ambio, 40*, 417–424.

Miesch, A. T. (1975). Variograms and variance components in geochemistry and ore evaluation. *Geological Society of America Memoir, 142*, 333–340.

Milne, A. E., Webster, R., & Lark, R. M. (2010). Spectral and wavelet analysis of gilgai patterns from air photography. *Australian Journal of Soil Research, 48*, 309–325.

Minasny, B., & McBratney, A. B. (2007). Spatial prediction of soil properties using EBLUP with the Matérn covariance function. *Geoderma, 140*, 324–336.

Oliver, M. A. (Ed.). (2010). *Geostatistical applications for precision agriculture*. Dordrecht: Springer.

Oliver, M. A., & Carroll, Z. L. (2004). *Description of Spatial Variation in Soil to Optimize Cereal management*. Project Report No. 330. HGCA, London.

Oliver, M. A., & Webster, R. (1987). The elucidation of soil pattern in the Wyre Forest of the West Midlands, England. II. Spatial distribution. *Journal of Soil Science, 38*, 293–307.

Oliver, M. A., & Webster, R. (2014). A tutorial guide to geostatistics: Computing and modelling variograms and kriging. *Catena, 113*, 56–69.

Papritz, A., & Moyeed, R. A. (1999). Linear and non-linear kriging methods. Chapter 18 in V. Barnett, A. Stein & K. Turkman (Eds.), *Statistics for the environment 4: Statistical aspects of health and the environment* (pp. 303–336). Chichester: Wiley.

Petitgas, P. (1993). Geostatistics for fish stock assessments—A review and an acoustic application. *ICES Journal of Marine Science, 50*, 285–298.

Petitgas, P. (2001). Geostatistics for fisheries survey design and stock assessment: Models, variances and applications. *Fish and Fisheries, 2*, 231–249.

Rousseeuw, P. J., & Croux, C. (1992). Explicit scale estimators with high breakdown point. In T. Dodge (Ed.), L_1 *Statistical analyses and related methods* (pp. 77–92). Amsterdam: North-Holland.

Snedecor, G. W., & Cochran, W. G. (1967). *Statistical methods* (6th ed.). Ames, Iowa: Iowa State University Press.

Sollitto, D., Romic, M., Castrignanò, A., Romic, D., & Bakic, H. (2009). Assessing heavy metal contamination is soils of the Zagreb region (Northwest Croatia) using multivariate geostatistics. *Catena, 80*, 182–194.

Soares, A. (2010). Geostatistical methods for polluted sites characterization. In P. M. Atkinson & C. D. Lloyd (Eds.), *GeoEnv VII—Geostatistics for environmental applications* (pp. 187–198). Dordrecht: Springer.

Stein, M. L. (1999). *Interpolation of spatial data: Some theory for kriging*. New York: Springer-Verlag.

Thiessen, A. H. (1911). Precipitation averages for large areas. *Monthly Weather Review, 39*, 1082–1084.

Vázquez de la Cueva, A., Marchant, B. P., Quintana, J. R., de Santiago, A., Lafuente, A. L., & Webster, R. (2014). Spatial variation of trace elements in the peri-urban soil of Madrid. *Journal of Soils and Sediments, 14*, 78–88.

Webster, R. (2000). Is soil variation random? *Geoderma, 97*, 149–163.

Webster, R. (2010). Weeds, worms and geostatistics. In M. A. Oliver (Ed.), *Geostatistical applications for precision agriculture* (pp. 221–241). Dordrecht: Springer.

Webster, R., & McBratney, A. B. (1987). Mapping soil fertility at Broom's Barn by simple kriging. *Journal of the Science of Food and Agriculture, 38*, 97–115.

Webster, R., & Boag, B. (1992). Geostatistical analysis of cyst nematodes in soil. *Journal of Soil Science, 43*, 583–595.

Webster, R., & Oliver, M. A. (1992). Sample adequately to estimate variograms of soil properties. *Journal of Soil Science, 40*, 493–496.

Webster, R., & Oliver, M. A. (2007). *Geostatistics for environmental scientists* (2nd ed.). Chichester: Wiley.

Webster, R., & Lark, R. M. (2013). *Field sampling for environmental science and management*. London: Routledge.

Webster, R., Welham, S. J., Potts, J. M., & Oliver, M. A. (2006). Estimating the spatial scales of regionalized variables by nested sampling, hierarchical analysis of variance and residual maximum likelihood. *Computers & Geosciences, 32*, 1320–1333.

Wiener, N. (1949). *Extrapolation, interpolation and smoothing of stationary time series*. Cambridge, MA: MIT Press.

Wold, H. (1938). *A study in the analysis of stationary time series*. Uppsala: Almqvist and Wiksells.

Youden, W. J., & Mehlich, A. (1937). Selection of efficient methods for soil sampling. *Contributions of the Boyce Thompson Institute for Plant Research, 9*, 59–70.

Zhong, B., Liang, T., & Li, K. (2014). Applications of stochastic models and geostatistical analyses to study sources and spatial patterns of soil heavy metals in a metalliferous industrial district of China. *Science of the Total Environment, 490*, 422–434.

Index

© The Author(s) 2015
M.A. Oliver and R. Webster, *Basic Steps in Geostatistics: The Variogram
and Kriging*, SpringerBriefs in Agriculture, DOI 10.1007/978-3-319-15865-5